CIVIL ENGINEERING SYSTEMS ANALYSIS

CIVIL ENGINEERING SYSTEMS ANALYSIS

LUIS AMADOR-JIMENEZ
Concordia University, Canada

CRC Press is an imprint of the
Taylor & Francis Group, an **informa** business

MATLAB® is a trademark of The MathWorks, Inc. and is used with permission. The MathWorks does not warrant the accuracy of the text or exercises in this book. This book's use or discussion of MATLAB® software or related products does not constitute endorsement or sponsorship by The MathWorks of a particular pedagogical approach or particular use of the MATLAB® software.

CRC Press
Taylor & Francis Group
6000 Broken Sound Parkway NW, Suite 300
Boca Raton, FL 33487-2742

Printed on acid-free paper
Version Date: 20160419

International Standard Book Number-13: 978-1-4822-6079-3 (Paperback)

Library of Congress Cataloging-in-Publication Data

Names: Amador-Jimenez, Luis, author.
Title: Civil engineering systems analysis / Luis Amador-Jimenez.
Description: Boca Raton : Taylor & Francis, CRC Press, 2017.
Identifiers: LCCN 2016017853 | ISBN 9781482260793 (pbk.)
Subjects: LCSH: Civil engineering.
Classification: LCC TA147 .A43 2017 | DDC 624.01/1--dc23
LC record available at https://lccn.loc.gov/2016017853

Visit the Taylor & Francis Web site at
http://www.taylorandfrancis.com

and the CRC Press Web site at
http://www.crcpress.com

Printed and bound in the United States of America by Publishers Graphics, LLC on sustainably sourced paper.

In loving memory of my dad (Luis Amador Sibaja) and

my grandmothers Claudia and Marina Chavarría Robles.

Contents

List of Figures

List of Tables

Preface: An Optimized Life

Individuals make hundreds of decisions every day, sometimes consciously but often unconciously. In doing so, we aim to achieve certain goals but, however, face restrictions. By reading this book, you aim to learn, or perhaps simply to obtain a good grade at school, but you are faced with a limited amount of time available to dedicate to it. A few personal decisions are transcendent enough to be organized into a more formal framework. Those consuming significant amounts of their own resources (purchase of a house/car or a major trip overseas) surpass a threshold level that force us to look into alternatives and to choose carefully.

This book deals with the methods and techniques that a civil engineer can use when analysing a system. It provides a means for supporting the decision-making process for the allocation of resources under circumstances with either conflicting goals or limited availability. The book presents two types of chapters; those intended to provide the basic foundations in mathematics, statistics and economics, and those that introduce and develop the application of the methods to real-world scenarios.

Civil Engineering Systems Analysis is a textbook of reference that teaches how to analyse engineering problems; however, its models are limited to the variables and facts incorporated into the analysis. Other variables and facts remaining outside the analysis should be used in a secondary stage to further reduce the set of choices or to choose the most convenient alternative.

The book is structured as follows: Chapter 1 illustrates how to take a problem, identify its nature and obtain a mathematical representation. Chapter 2 introduces some concepts from mathematical analysis required to define the scope of applicability of problems appearing in more advanced chapters or to provide a means to warrant the existence of a solution. It creates a strong foundation that is typically missing in most civil engineering programs of study. Chapter 3 presents some concepts of optimization. After learning how to take a problem and enunciate it in Chapter 1, Chapter 3 teaches you how to approach and solve the mathematical formulation. To make it simple, Chapter 3 looks at basic problems of decision-making and methods to select the best solution. Chapter 4 covers concepts you may have learned before in probability and statistics. Chapter 5 extends the concepts of Chapter 4 into estimation and prediction. Chapter 6 addresses a special type of model for the integrated analysis of land development and trips on a network originated by travellers or the movement of freight. Chapter 7 covers problems of transportation and municipal engineering. Chapter 8 applies the learnings to civil infrastructure management. Finally, Chapter 9 introduces uncertainty through decision trees and the use of conditional probabilities in decision-making.

Before we jump into that, I would like to close this preface by encouraging you to take the methods presented in this book and attempt to apply them as you learn them. Think about common day-to-day problems: Have you ever walked into the metro platform to a position that perfectly aligns you to the exit at your exit stop while you wait for the train to arrive? Do you ever purchase at the grocery store by analysing calories, fat and nutrient contents in addition to the price? Or do you ever buy what is on sale and enough to last until the next time it will be on sale? As you will soon notice, eventually, any decision you make in your personal and professional life can be optimized.

Luis Amador-Jimenez

1

Introduction to Modelling

Civil engineers deal with three main tasks during their career: analysis, design and construction. When analysing, engineers are commonly required to verify and suggest means to comply with regulations or ways to improve the performance of a system, and, in doing so, they seek to achieve performance goals. In designing, engineers face the need to create built environment, and this typically results in conflicting objectives of professionalism, developer's interests and resources. Finally, during construction, an engineer faces the need to allocate labour, equipment and materials to reach given milestones on time. However, timely achievement of such deadlines requires more intensive use of the resources mentioned which triggers project cost. Hence, the engineer is faced with two common conflicting goals: minimization of resource utilization and achievement of a certain level of quality and/or performance.

1.1 Decision Elements

This morning, when you woke up, did the possibility to continue sleeping cross your mind? If so, then you had to make your first decision of the day. Why do you like to sleep? Your body gets some rest, you recover your energy and even your capability to learn reaches its highest point. So why then did you stand up and initiate your day instead of sleeping? You possibly either had breakfast, took a shower, exercised or finished your video game. Some time ago, economies defined a concept called utility, for you and me (who stay in bed for another 30 minutes). Sleeping gives us satisfaction, is valuable, is useful or, in other words, gives us utility. But the utility of sleeping was not as important for those who woke up and started their daily activities, perhaps because they have already slept for over 10 hours.

The amount of utility one gets out of sleeping decreases with time: sleeping after a long day of work/study has a very large utility, sleeping more after the first hour of dreaming has a large value (utility), sleeping after having dreamt for 6 hours still has some value (utility) and sleeping after 10 hours possibly has no value. This behaviour of decreasing value is formally called decreasing returns to scale: the more you get of something, the less you value it. Think of it as the amount of your favourite beverage: the first glass is amazing, the second glass is refreshing, the third glass is fine and the fourth glass is not so good.

Many other activities are valuable to an individual and provide him/her with utility. Eating, socializing and sleeping are just a few examples. An individual consumes food items to eat and in doing so has his or her own preferences: he may not like (and/or never) eat pork but love chocolate. In socializing, an individual perceives value out of activities and items: silver/gold jewellery, brand watches, fancy cars, visiting clubs, organizing parties at his place, attending conferences and seminars or publishing research in prestigious journals. All these activities require the consumption of goods. Goods cost money and money is in short supply (ask your parents if you do not believe me).

On a daily basis, individuals attempt to achieve the highest possible value of utility, and for that, they dedicate time and efforts to achieve what they want and are required to expend money. The problem of any individual is to maximize his utility given a certain amount of resources (time, money, ability).

A similar problem faces the junior engineer at my consulting company when I ask her to take care of the entire design process for a new building, including obtaining construction permits. She starts by meeting with the customer to learn his office space/functionality needs, then she has to conceptually design a building layout and validate the design with the customer. Assuming the client is happy, she will order the structural, mechanical and electrical engineers to design the corresponding components of the building. After a month or so, the blueprints will be ready and she will have to visit the municipality and other government institutions to obtain the required construction permits.

Why does she do all this? It sounds after all complicated and stressful. She perceives utility from the salary and the working experience. The company perceives utility out of the revenue (fees charged to the customer) minus the expenses we incur in (including her salary). Resources are involved in these activities and act as constraints: the company uses time from drawing technicians, structural, mechanical and electrical engineers; paper for the blueprints; computers that deteriorate for the designs and calculations; electricity; and pay rent for the building, among many others. The problem of any firm is to decide how much resources to use in order to spend as little as possible but yet to be able to satisfy its customers and obtain large amounts of profits.

The problems we have presented so far are given at a moment in time. The individual wants to achieve a large amount of utility today, and the company seek profits this month. Time complicates things and makes the problem more interesting (I bet you differ). The individual not only faces a maximization of utility today but for the rest of his life. Tomorrow and the day after and next year and in 5 years and in 20 years, he will wake up and will face a similar problem and will count with similar restrictions.

If I go to the cinemas twice a month, travel to sun destinations once every 2 months and attend one conference per year, I would like that to remain the same across time or to improve. Would you not? You eat your favourite meal once a week, purchase a new pair of shoes every month and go out with your friends every Friday. Would you not like being able to do any of those next year? Individuals get comfort out of the levels of consumption across time; they seek to have constant levels of consumption. In economic theory, this is called smooth consumption. In the same manner a company wishes to obtain every year similar amounts of profits; actually if you ask a financial planner, he will tell you the company seeks increasing (non-declining) levels of profits.

1.2 Objectives and Constraints

Seeking the largest amount of utility or profits is called *maximization*, and seeking the usage of the smallest amount of resources is called *minimization*. The amount to be minimized or maximized is called an *objective*. Restricting yourself to a given amount of some resource gives you a *constraint*. The amount of resources you will use is called decision variable or control variable. How effective you use your time, your machines, etc., gives you *parameters* which are typically related to productivity. Let us look at some examples.

Constraints impose limits to the ability of the decision maker, and they restrict the 'universe' of choices by considering what is feasible; for instance, producing a very large amount of ice cream may not be feasible because of storage limitations. Constraints can also limit our ability to combine resources, sometimes to produce goods or build, others to achieve intermediate elements.

The space of choices as delimited by the constraints is called feasible space. In general, one requires a 'well-behaved' feasible space as a requisite to solve an optimization problem and find a solution. What I mean by that will be the matter of another chapter. For now, let's allow a loose usage of objective, constraints and decision variables.

1.3 Identifying Objectives and Constraints

One of the main issues when we deal with a problem is to narrow it down to its bare bones and capture its very nature. A model is after all a representation of reality that have abstracted many facts, concentrating only on the ones that matter for the decision maker. As a civil engineer, you will have to identify the objective as the most important element of your problem. Consider, for instance, a frequent flyer who takes flights every week but pays from his pocket.

Such traveller may wish to have a short flight, such that he arrives at his destination on time; he may want to be served free-of-charge meals or to be able to use an in-flight entertainment system with a good selection of free movies, TV shows, magazines, news, music and games. But if he pays from his pocket, the most important element for him will be the cost, how much he will have to pay overall for every trip, and hence, the objective will be the minimization of the cost.

The other elements will become constraints: the traveller may wish to have a minimum number of free films (say, one film for every 2 hours of flight time) and a minimum number of foods (say, one complete meal every 5 hours of flight). And possibly he may want to limit his options to no more than one stopover for every 6 hours of flight. All these elements turn out to be constraints. A rather different problem faces the top executive whose company pays for the flight. He does not care that much for the price of the flight; he cares for his comfort. Hence, he will try to maximize the number of amenities he counts with (number of features or conveniences). The constraint is perhaps not to expend more than the budget that the company provides him with for every trip.

The lesson learned is that an objective is nothing but a constraint without limitation, so objectives and constraints can be interchangeably used, but duplication should be avoided. Typically, we define only one objective and many constraints. In more advanced problems, one can have many objectives and even define a problem only in terms of multiple (conflicting) objectives.

One last word about constraints, sometimes we forget about evident constraints, such as the impossibility to use negative values (non-negativity), or the limitation of one choice every period. These types of constraints must also be defined. Many times we use some 'black-box' software to solve a problem for us, and as such having a correct problem definition results essential.

1.4 Creating a Decision Model

Building a model requires three major steps: (1) an identification of the very nature of the problem, (2) writing a mathematical formulation and (3) solving it. There is, however, not a given formula to do this. Understanding a problem requires us to identify the features that we have control over and that by manipulating could help us achieve a goal. Problems typically involve multiple dimensions and considerations of other systems commonly need to be removed or simplified in many cases.

1.4.1 Identifying the Nature of the Problem

If possible, start by constructing a sketch, scheme, diagram, table or any other visual representation of the problem and its mechanisms. This informal tool will definitely help you understand the problem's nature. For any problem, look at its elements; write them down in a blank piece of paper, one by one, and use arrows to connect them and brackets to group them. Write them all down and somehow connect them.

Continue by exhaustively enlisting possible combinations of resources and how they may be limited; this will lead you to the constraints. More than that, it will allow you to identify decision variables. When writing constraints, bare in mind that constraints can be easily turn into objectives: so whenever you don't see a limit, but there is a way to combine elements, you are possibly on the presence of an objective.

Enlist exhaustively all constraints and choose if possible one as your objective; if not, attempt to assign a relative importance to each on any scale (say 1–10). Once you have done so, assign letters to the elements that you have combined; use letters that make sense or simply use x with a subscript x_i to differentiate among them. If more dimensions are involved (time, weather, alternative, origin, destination, etc.), expand the subscript to include this; the use of superscript is preferably reserved to cases where the quick visual distinction of the dimension being indexed really matters during the analysis.

1.4.2 Constructing a Mathematical Formulation

It is time now to formally express the problem. Start always with the objective. You have two choices: *MAX* for maximization or *MIN* for minimization. That depends on what your goal is. Underneath the objective, write the letter(s) that corresponds to the decision variable(s). Then continue by writing the following words: *subject to* or simply *s.t.*, after which you will enlist the constraints including non-negativity and other logical constraints.

Let us look at a couple of simple examples. Imagine you are a town planner who just learnt that the federal and provincial government will fund 2/3 of a new pipeline project plus a water tank in order to secure water supply to your town. Records of water consumption for the entire town every month are available and based on previous observed history as shown on Table 1.1.

Population is stagnant (no growth) for the past 20 years. The federal government already established as a condition for funding that you should at least use a pipe such that you directly get in town a flow equivalent to the month with least requirement. What elements would you be interested on? Perhaps the capacity of the pipe and the tank? Now, how would you create the problem? Let us see: take a piece of paper and write the information that is given to you, and attempt to connect it.

TABLE 1.1

Water Consumption Demand

Month	Consumption ($\times$1000 m^3)
January	72
February	73
March	75
April	74
May	77
June	112
July	120
August	145
September	110
October	80
November	75
December	73

TABLE 1.2

Water Consumption: A First Attempt at Solving

Month	Consumption ($\times$1000 m^3)	Tank Capacity
January	72	0
February	73	1
March	75	3
April	74	2
May	77	5
June	112	40
July	120	48
August	145	73
September	110	38
October	80	8
November	75	3
December	73	1

Perhaps some of your initial thoughts are whether to use the minimum of 72 as your pipe supply. You could even attempt to solve manually for the tank capacity. Table 1.2 is only intended to illustrate such approach.

According to this first approximation of our thoughts, we will need a water tank of 73 ($\times$1000 m^3). Although incorrect, this first approximation to the problem has taught us that we may well use a number such that the water tank never runs dry. Can we use the mid distance between lowest and highest demands, that is, average demand? Should this reduce to the minimum the

TABLE 1.3

Pipe Capacity as the Least-Square Difference

Month	Demand (×1000 m^3)	Pipe Supply	Monthly Deficit
January	72	90.5	18.5
February	73	90.5	17.5
March	75	90.5	15.5
April	74	90.5	16.5
May	77	90.5	13.5
June	112	90.5	−21.5
July	120	90.5	−29.5
August	145	90.5	−54.5
September	110	90.5	−19.5
October	80	90.5	10.5
November	75	90.5	15.5
December	73	90.5	17.5

tank capacity requirement? If you do so, you will find yourself in the position of Table 1.3. Most importantly, we have already identified that the objective is to minimize the monthly difference between demand and supply.

Even more our decision variable is such supply; let us call the pipe supply x and each month demand D_i. So we want to minimize $x - D_i$; as a matter of fact, this minimization is for every month, so we will write a summation and use i to represent the month, i will go from 1 to $n = 12$.

Additionally, note that this difference could give us positive or negative values, and at some point, they may cancel each other; so it is rather better to use the square of the difference to avoid this situation. This lead us to the following formulation.

$$MIN \sum_{i=1}^{n} (x - D_i)^2.$$

The decision variable is x, and as we learnt from calculus, it suffices to take the first derivative of this expression with respect to x to obtain an answer which would be $x = 90.5$. If we add the values of column 4 of Table 1.3, we find that they add to zero; hence, we have truly minimized the difference.

Having found that 90.5 is the pipe capacity solves half of the problem. What is the tank's required capacity? Is it 54.5 (i.e. the maximum deficit for the month of August)? From January to May there is an excess of water we could accumulate and start using it on June when the pipe does not provide us with enough water to cover the town's needs (demand). If we do so, we would need to look at the monthly cumulative as shown in Table 1.4.

TABLE 1.4
Tank Capacity as the Maximum Deficit

Month	Demand ($\times$1000 m^3)	Pipe Supply	Cumulative Deficit
January	72	90.5	18.5
February	73	90.5	36
March	75	90.5	51.5
April	74	90.5	68
May	77	90.5	81.5
June	112	90.5	60
July	120	90.5	30.5
August	145	90.5	−24
September	110	90.5	−43.5
October	80	90.5	−33
November	75	90.5	−17.5
December	73	90.5	0

So you would have a tank of 43.5 ($\times$1000 m^3). If you select such capacity, you may be failing to consider how those elements from the real world that we neglected at the beginning of the problem may affect the required capacity of the tank.

We are making a huge investment on the premise that demand would not grow and that there are no losses of water on the system. What about pressure/pumping stations? Or simply the fact that we are only looking at monthly averages? And the daily demand could turn our results invalid. At least, we have been able to get a working model that we can now expand to incorporate the other factors.

Finally, before ending with this problem, let us look at the feasible space. What values of the decision variable can we take? As defined here the pipe supply can go from zero to any number; hence, there is no upper limit; hence is open (not closed), look the decision variable although continuous goes from (0,...) and the problem is unbounded, hence is not compact.

Thanks to the fact that the objective function moves in the direction of minimization, we can obtain an answer. If under any circumstance you were looking for the *MAX*, there would be no answer. We could, however, expand the problem and bring a monetary budget limitation, or a pumping capacity cap, and hence make it feasible by establishing an upper bound.

1.5 An Overarching Problem

We will define here a problem that will inspire many sub-problems used through the text. We will start by attempting to define the decision-making world of civil engineers during their professional life. A civil engineer is faced with either the design, construction or retrofit of infrastructure and buildings. If we look at buildings, the engineer will look at the use of the building and categorize it as residential, commercial, industrial, etc. For the design, the engineer faces the need to minimize the size of the foundation and the structure while restricted to standards specified by norms and codes. In addition, the engineer will look at the purpose of the building and attempt to minimize the amount of space while still complying with the number of parking spaces, offices, bathrooms and other elements as required by urban planning regulations and building by-laws.

When the engineer is tasked with the design of infrastructure, he will have to look at either surface or underground infrastructure and decide on the dimensions and features required as preconditioned by the demand (usage) from the public at large. For these, the engineer may need to play the role of planner and forecast demand for the infrastructure. Later on, the engineer will need to worry about how to maintain such infrastructure during its lifespan (Chapter 8).

Let us now set ourselves on the feet of a planner attempting to analyse the world of surface transportation demand and its links to urban planning and economic environment. This will be further explored in Chapter 6.

For simplicity, we will define a downtown area with suburbs and a rural region on the outskirts of our land use fabric. Land uses play an important role in defining the space of the problem; now is the time to bring the users. The decision makers will follow a positive self-selection process in which they will choose their location based on convenience; for per instance, housing will develop on the suburbs similar to retail stores. Government offices, service industries and company headquarters will locate in the downtown area. The labour market (i.e. employment or jobs) will hence concentrate there, but the other industries will attract some employment as well. In the rural area, we find basic industries (Chapter 6): agriculture, energy, forestry and raw materials. The reason behind this is the availability of such resources in such areas (Figure 1.1).

A flow of people (called commuters) will travel every day in the morning from their residences to the labour market and return back home in the afternoon. This defines the peak hours of the day. And for the civil engineer, this means the demand on mass transportation modes (trains, metro, buses, tramway, walk, bike and cars). Figure 1.2 illustrates the flow of commuting trips from housing to jobs.

A flow of commodities (goods) will be to move from the basic industries (energy, raw materials, wood, agriculture) to intermediate industries that will transform them into final goods. Energy will move as well but through wires (same as telephones and cable services such as the Internet). Figure 1.3 illustrates such flow.

A final movement of merchandises will flow from intermediate industries to final consumption markets (Figure 1.4). In this case, this is within the

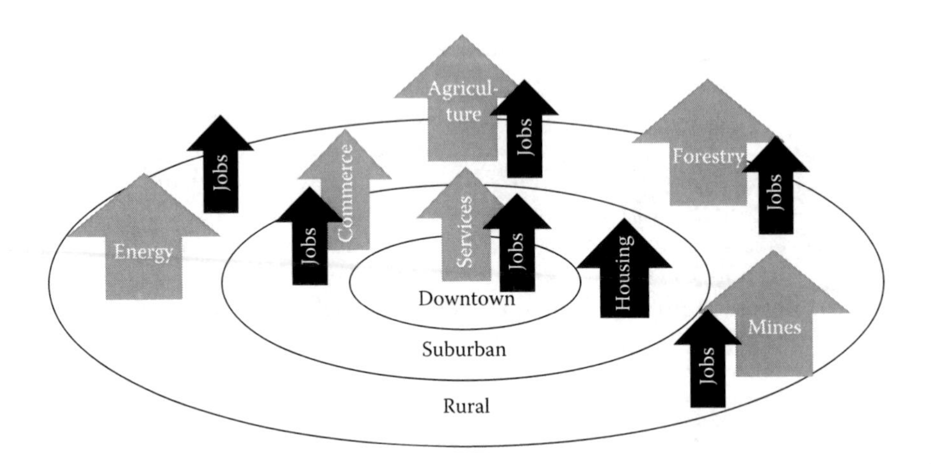

FIGURE 1.1
A basic land use location set-up.

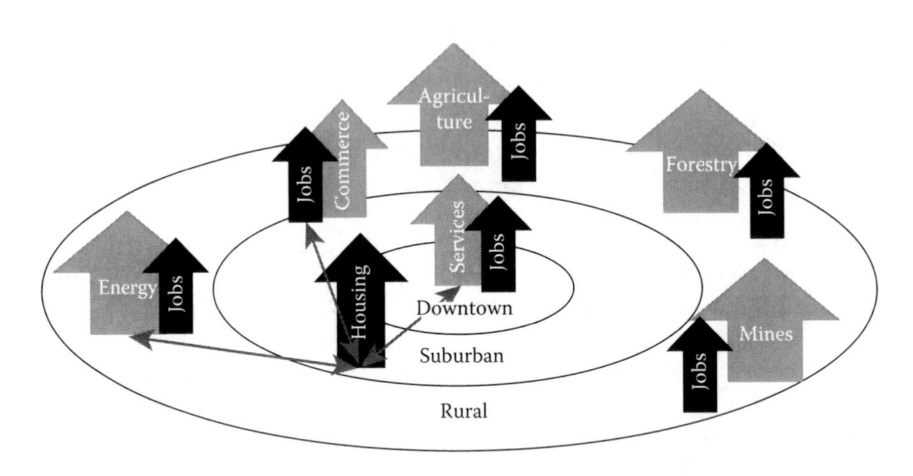

FIGURE 1.2
Commuting trips.

region or to other regions. A warning seems necessary, as suggested before the model has oversimplified reality and prevented the movement of raw materials to other regions.

The condition that all raw materials will be transformed by local companies is one of such abstractions the decision maker or the planner will impose to simplify things, but that should be invoked back at the end when the analysis needs to be reconsidered in light of such limitations and simplifications.

It is always useful to explicitly state the assumptions of the model in a list format. For this example, we have several assumptions summarized in Table 1.5.

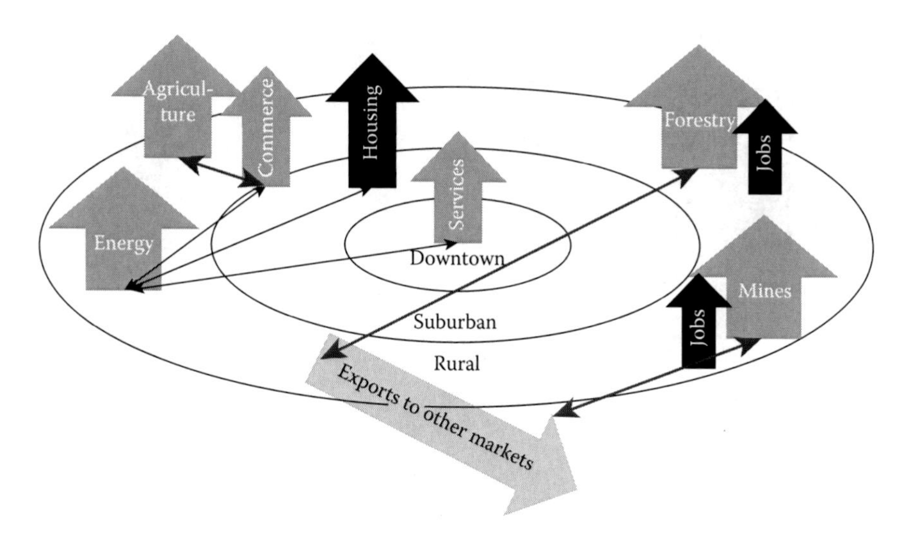

FIGURE 1.3
Freight trips to intermediate producers.

Now that the model and its environment have been defined, it is time to define the decision variable. If say a planner is looking to programme bus trips between two given points of a downtown area, she will be required to get the demand for trips between such given points. Call x your decision variable and use i to denote the origin and j to denote your destination; your decision variable turns out to be x_{ij}. To specify that the trips belong to passenger cars (c) and opposed to freight (f), we could use a superscript such that x_{ij}^{c}. If you care about time and want to include it as a dimension, then perhaps you can simply expand the subscript dimension and have a third index.

An objective for such planner is to have the minimum number of bus trips on the network as they cost money. This will be constrained by available budget and maximum level of total green house gas emissions (GHGs) among others.

For freight trips, the planner may be more concerned about the deterioration of the roads from the number of trips that tractor trailers moving freight produce. The number of trips is out of the control of the planner (as opposed to the allocation of buses which are commonly operated by the city). A possible objective for the planner is to maximize road condition by allocating preventive interventions (overlays, crack-sealing, etc.). One possible constraint is the annual budget. The problem could be turned around, and the cost change into an objective, meanwhile the road network condition could be transformed into a constraint by requiring a minimum average network level of condition every year for the coming decade.

A decision maker who has analysed the problem and obtained a solution should now consider how his simplification of the world may affect

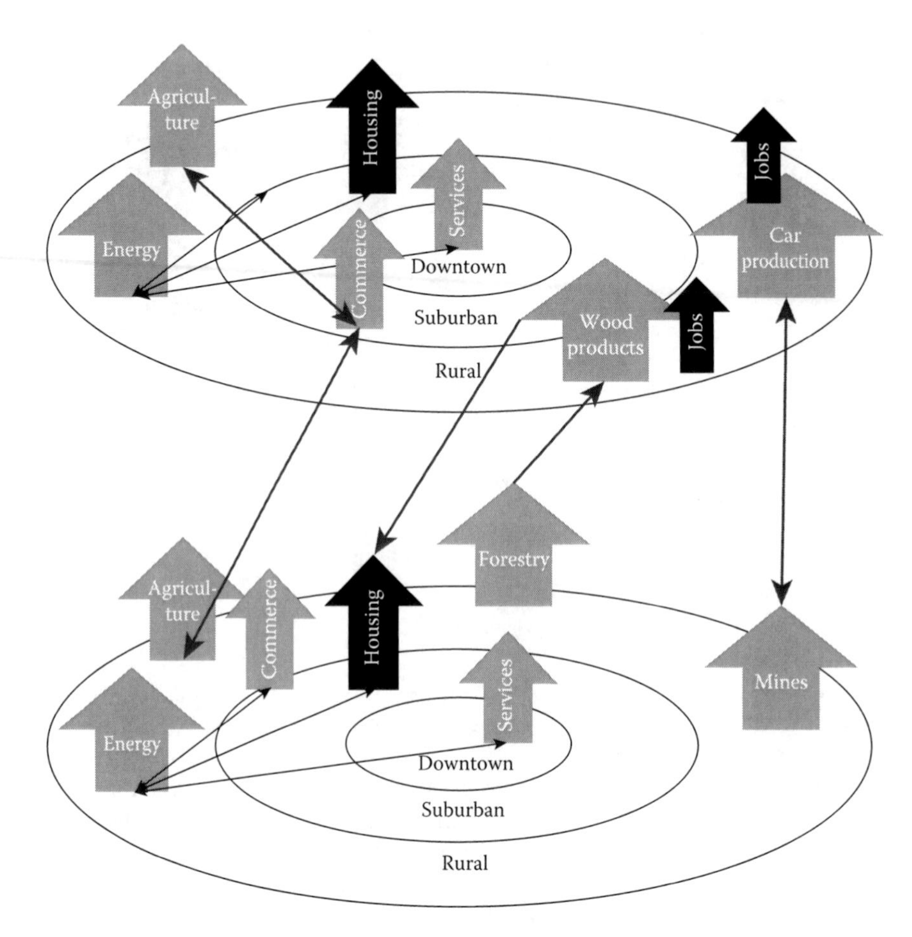

FIGURE 1.4
Freight trips to intermediate producers on other regions.

TABLE 1.5
Explanation of Trips' Purpose or Nature

Assumption	Detailed Explanation
1	All commuting trips from suburbs to downtown
2	All raw materials to be transformed within the region
3	Intermediate products produce intra- and inter-region trips
4	All industries of the same type aggregated to one big company
5	Aggregation of trips from one zone to the other

his choice of answers. In this sense, the planner that identified 100 million dollars per year as required expenditure to achieve required levels of service must consider that 1% of the raw materials are moved outside the region, and should adjust his model accordingly (possibly the adjustment would not make a difference), but if the percentage is large (say 20%), then the consequences should be considered and the model fine-tuned. How to do that will be discussed in the subsequent chapters.

Exercises

1. ConcreteMixer, Inc. was contracted by Infradevelopper, Inc. to produce Portland concrete class III, which is a high-resistance waterproof concrete. What are the objective(s) and constraint(s) of each company?
2. A steel re-bar company is contracted by InfrasRoad, Inc. to produce 100,000 re-bars, type 45M of 6 m long. For this task, they need to use their machines, raw iron and labour. What is/are the objective(s)? Constraints?
3. You are hired to design a house for a wealthy customer. He wants 10 rooms, 10 bathrooms, a cinema room, a gym, an indoor pool, a garage for 15 cars, 2 kitchens and at least 3 terraces. The house must look like a castle from the outside but be modern from the inside. He is not concerned about the cost, but cares about having a smart home with voice control for blinds, lights, temperature control, ambient music and electric devices. Controlling each possible aspect is requested. What is your objective? Constraint(s)?
4. A structural engineer is asked to design the new *Champlain Bridge* connecting *Montreal* and *Brossard* regions. What could be the objective and constraints?
5. A transportation engineer is asked to design the new *Champlain Bridge*. What could be the objective and constraints?
6. A road maintenance engineer is asked to reduce the amount of gas emissions produced by the annual maintenance and rehabilitation of the pavements within a metropolitan area. For this, the planner has some control over the maintenance and rehabilitation treatments employed to rehabilitate pavements. What could be the objective and the constraints?
7. A water planner wants to make sure there is enough water coming to town through the pipeline that feeds in the river and brings water into the water treatment plant. What would be his objective?

   Now, try to think what elements would condition or constrain this objective that actually push the decision maker to move in the direction contrary to that of the objective.

8. What is the objective of a structural engineer dimensioning beams and columns for a high-rise building? Which ones would be his or her constraints? Think of the fact that this engineer follows codes such as the building codes and the seismic codes, which guide the designer to allocate certain amounts of reinforcement to the concrete elements and to use certain strength for the concrete itself.
9. Consider the position of a manager who is in charge of making sure there is no flooding in town; hence, the manager must replace pipes in order to increase their capacity and make sure there is no standing water during a period of intense rain. What is the objective and what are the constraints?
10. What is the objective of a project manager in charge of the construction of a new building in downtown Montreal? Formulate the objective and the constraints and then modify your formulation by converting a constraint in the new objective and the old objective in a new constraint.
11. An engineer is being asked to design a dam for a hydroelectric company. What would be the various possible combinations of objectives and constraints?
12. The city hall has received thousands of complaints regarding the improper functioning of the mass transportation system (buses, metro, trains) in the city, and 90% of the people complaining are travellers going to work, while the rest are students going to school. What should be the objective and what constraints can you identify?
13. A student is trying to pass a certain class at school. What is her objective and her constraints?
14. A foundation engineer is hired to conduct soil explorations in order to design the foundations of a building. The idea is to have a good idea of the bearing capacity of the soil; the more exploration the engineer conducts, the more accurate will be the characterization of the bearing capacity. However, this is also expensive and time consuming. Identify possible objective-constraint pairs for this problem.
15. As airport manager you perceive that the airport needs to go through a rehabilitation. This requires soil exploration to determine where the weak segments of the pavement are. This also requires a measurement of the thickness of the pavement in order to make sure the strength is adequate.

    These two elements can be tackled through sampling with boreholes. Finally, you need to measure the surface skid resistance to ensure safety. You know your maximum budget is one million dollars for getting the job done. What could be your objective and constraints?
16. You have to travel to Miami for a conference. The university gives you $1000 for such purpose. Enunciate your objective and constraints.

17. Consider now a business executive that needs to travel every week. Given the company pays for his plane ticket, what would be possible objectives and constraints?
18. Let us twist the previous example and let us think of the poor student trying to travel home for vacations. She pays for everything from her pocket. What would be her objective and constraints?
19. What is the objective of a driver whose company pays for the rental, the fuel and the food involved in a road trip?
20. What is the objective and constraints of a driver using his own car to travel during the long weekend for vacations to New York?

Answers

1. Holcim's objective is to minimize cost and constraint is that concrete characteristics are adequate. SNC Lavalin's objective is to acquire the maximum concrete strength and waterproof possible while not spending more than the given budget.
2. *Possible objectives*: Minimize cost of production or maximize productivity. *Constraints*: Number of re-bars each machine produces per hour, number of hours employees work per day (or per week).
3. As a professional engineer, you must satisfy the customer's expectations. So the objective is to maximize number/capability of technological features. The constraint is the type of technology that is able to meet the client's requirements. Budget is not a constraint in this case as the customer is a rich millionaire who does not care!
4. *Objectives*: Minimize cost or minimize re-bar size and amount of concrete. *Constraint*: Achieve sufficient structural capacity to comply with steel/concrete codes.
5. *Objective*: Maximize the bridge's traffic capacity (i.e. number of vehicles that can use the bridge). *Constraint*: Budget.
6. The objective is to minimize the total annual amount of gas emissions (or you can also do it for another period of time such as a semester or a month, but the typical choice of a year is to match the annual operation and budget cycle).

   The constraints are the level of condition and the overall quality of the roads. As you can imagine without this constraint, the optimal solution would be to do no maintenance and rehabilitation activities because that results in zero gas emission.

   We need a minimum level of quality on the condition of the roads to push in the opposite direction to the minimization of gas emissions (i.e. minimization of rehabilitation and maintenance treatments).

7. The objective is to maximize the amount of water coming into the water treatment plant. The problem is the constraints. For instance, there is a budget that will restrict the ability to have a very big pipe, so pipe diameter will be conditioned by budget: the bigger the pipe, the more expensive it is.

   Another constraint comes from the fact that the water treatment plant has a restricted capacity as well. So it is pointless to have a lot of water if it cannot be treated.

8. The engineer will try to have a design that complies with codes; codes commonly provide you with minimum strength, dimensions, separation, reinforcement (re-bars), etc.

   The problem is that the customer has a budget and often there is a conflict between the client and the designer because the design implies a large amount of money. So the objective is to have a safe design and to maximize the structural adequacy of the building.

   The constraints are not only the budget but also the minimums required by the code. This puts the engineer in a complicated situation wherein sometimes element spacing, type of materials and location of columns and beams may need to be revised in an attempt to reduce the cost while still warranting an adequate design.

9. The manager attempts to maximize total network capacity for the pipes in the system. This implies replacements, and replacements are limited by not only the budget, and they are also limited by the season of the year.

10. One of the many possible objectives is to minimize construction time; in this case, the budget and the quality of the construction will become constraints.

    Alternatively, one can use the cost as an objective and state a minimization of cost while constraint by a time budget (no more than 12 months for the completion per instance) and a quality of construction constraint (no less than 9 in overall quality score).

    One can also make the quality an objective and state a maximization of construction quality while restricted by budget and time constraints.

11. The dam must be safe, so you may want to maximize structural integrity while constrained by a money budget.

    This can be changed to minimize budget while constrained by a minimum level of safety on the design.

12. The objective for the city planner would be to minimize total travel time for all passengers on the network. The constraints in this case would be the total budget available for changes and the number of operators and vehicles available for the movement of people.

Another possible objective could be to minimize cost while constrained by a maximum travel time. The establishment of such travel time is a complicated matter we will try to tackle in Chapter 7.

13. The student may use as objective to maximize her grades and her constraints would be total time to study, total budget for books and other support materials.

A possible alternation would be to have as objective to minimize total cost and effort in time used to study, and in this case, the constraint will be to obtain at least a B-grade.

14. The foundation engineer's objective could be to maximize certainty for the characterization of the soil's bearing capacity and the constraint will be the budget available for soil exploration.

A possible objective could be to minimize cost while having 80% confidence that the soil characterization is correct.

15. The airport manager would want to minimize total cost, and in this case, the constraints will be to make sure the strength of the pavement is at least superior to some minimum value. The other constraint will come from the skid resistance of the surface to make sure you warrant certain level of safety for the planes that are landing or taking off.

A possible alternation of this objective is maximizing the strength of the pavement, and the skid resistance and budget become objectives. Alternatively, you can have maximizing skid resistance as objective while constrained by a money budget and minimum overall strength.

16. The objective is to minimize cost (because that means money in your pocket) the constraint is to arrive to your destination.

17. The business executive wants to maximize his comfort–defined as legroom, space for his own, amenities and features such as having a meal on the plane, enjoying entertainment through the availability of films, music, games, etc. In addition, the ability to have a flat-bed seat on long-haul flights, the space separation to the next passenger (unless he enjoys somebody's elbows over his seat space).

So in summary, the objective is to maximize comfort. The constraints are arriving at the destination within a reasonable amount of time and under the given budget given by the company to pay for the ticket.

18. The poor traveller who pays for her ticket wants to minimize cost; of course, she also enjoys the amenities of entertainment and meals and the comfort of the legroom, overall seat width and if she can get into a business class seat she will, but all these elements will be part of the constraints. She will establish the minimum required amount of features in that regard.

19. The traveller will rent the best car he can, so he will maximize the luxury of the vehicle while restrained by a certain budget.

20. The student travelling to New York on a road trip using his own vehicle wants to minimize travel cost, reducing fuel consumption and overall travel expenditures. This traveller will pack home-made sandwiches, roll the car downhill and will not stop to rest unless strictly necessary (consider when you stop, you expend more fuel given where you have to go through a cycle of lower shifts). The constraints will be the level of comfort; he is not going to travel without food or a place to sleep at night.

2

Mathematical Analysis

2.1 Fundamentals of Topology

This chapter aims to characterize what we mean by a 'well-behaved' feasible space and objective function. In order to do that, it introduces basic concepts of set theory and its applications to optimization.

As opposed to traditional textbooks in mathematics, the definitions herein provided do not follow a rigorous process and are intended to be as simple as possible to provide the engineer with a quick understanding of the concepts.

Additionally, this book provides examples from day-to-day practice of civil engineers in their professional activities. The reader should keep in mind that these examples will be revisited on more advance chapters.

2.1.1 Sets

A set is a group of elements that share at least one common characteristic. For instance, all members of the student association are part of a set, and their 'membership' is their common characteristic.

Subsets can be defined from a set; following the previous example, all freshmen (within the student association) are a subset of the previous set. Here, the reader realizes that it is convenient to give an easy-to-recall name to each set, so, for instance, the set of student members of the association can be called by the capital letter A, while freshmen could be called F. The indication that a set is part of a bigger set is given by $F \subset A$ and reads F is a subset of A.

There are some classical sets widely used in all fields of knowledge and their letter-names are somewhat standard. For instance, all positive integer numbers called 'natural' are represented by N. All integers, whether positive or negative, are called by the letter Z; all rational numbers, that is, those defined as the ratio between two integers, are called by the letter Q. The rational numbers are classically defined as follows: $Q = x/x = \frac{p}{q}, p, q \in Z, q \neq 0$.

As the reader can see, rational numbers can be defined replicating the previous definition with the addition of a constraint: the denominator cannot be zero, which makes sense, because otherwise the whole fraction will tend to infinity and infinity is undefined.

It should be noticed that Q does not comprise those numbers that failed to follow the previous definition. So any root that cannot be solved, and special numbers such as the base of natural logarithms, belong to the set called irrationals and are commonly represented by I. The union of irrationals and rationals, mathematically denoted as $R \cup I$, creates the bigger set we all know as real numbers R (to distinguish from imaginary numbers).

A construction firm seeking to maximize their utility will define the space of possible profits as the natural numbers N because decimals are not relevant for managers, neither are negative amounts. However, a firm may well incur in losses; therefore, the definition may need to be revisited and changed to the set of integers.

In general, sets are used to either define the space at which a solution can be found, delimiting the alternatives to those that are relevant, or to characterize the elements that will be used in the analysis. For instance, when looking at traffic flows on a highway system, the engineer may want to delimit its analysis to only positive volumes of vehicles, and because we only care about complete vehicles, the space for the analysis will be composed of zero or positive integers.

If we look at all those ramps that enter a highway before a given control point, we can define a set of ramps R that are located before a given milepost.

2.1.2 Empty Set, Unions, Intersections, Complements

The empty set is the one containing no elements; however, it is still a set; you can think of it as the club where no one can satisfy its entry requirements and is denoted $\emptyset$. In the previous example, if we define a set composed by those highway ramps having more than 1900 vehicles/hour (veh/hour), and at the time no one satisfies this requirement, that is, every ramp observed is less than 1800 veh/hour, then there exists an empty set, which is denoted $\exists\,\emptyset$.

Now let us create a group from those highways with high volume of traffic, such that peak hourly traffic flow is at least 1000 (veh/hour); if existing highway ramps are as given in Table 2.1, then the new set will have three elements $A1, A4, A5$, and we could denote this set by a letter, say H. A formal definition of these groups of highways with high volume is $H : A1, A4, A5$. We could create another set of ramps comprised of those with flow strictly smaller than 1000 veh/hour and call it L. Such set, L, is called the complement of H and is defined as the set of ramps that is not in H and is denoted H^c.

The union of H and L generates R and can be denoted HUL; notice the union is defined by the operator *AND* which means it is comprised of those elements that satisfy both the requirements of H and the requirements of L. In this case, the intersection of H and L has no elements, thereby satisfying the definition of $\emptyset$.

TABLE 2.1

Highway Ramps and Peak Traffic Volume

Ramp Number	Peak Traffic Flow	Flow Direction
$A1$	1100	Entry
$A2$	800	Entry
$A3$	600	Exit
$A4$	1300	Entry
$A5$	1800	Entry and exit
$A6$	900	Exit

2.1.3 Sets and Spaces

The distinction between a set and a space is given by the ability to measure a distance between any two elements of a set. Sets without this property cannot be defined as spaces. It is rather convenient to work with spaces.

However, in some cases, sets are composed of qualitative information, such as ranges of bearing capacity for the foundation soil (low, medium or high) or condition of road segments (good, fair, poor). In these cases, it is impossible to have a distance between what we call good road condition and fair road condition. Luckily often we have a numerical characterization, and instead of ranges of bearing capacity, we may have values of California bearing ratio (CBR) such as 80, 53 and 25.

In this case, it is simple and easy to know the distance as it is based on the absolute value of the subtraction $|80 - 53|$. In two or more dimensions, it is simply given by the norm, represented by the square root of the summation of the square of the differences, that is, $\sqrt[1/2]{(x_i - y_i)^2}$.

2.1.4 Upper and Lower Bounds

In defining civil engineering problems, it is important to be able to quickly realize if there is an upper (lower) bound for the problem at hand. Boundaries play an important role in warranting that the problem space is not undefined. The first thing we need to consider is if the space is ordered somehow, typically by the smaller $<$ or bigger $>$ operators.

Existence of a smaller (largest) element implies a lower boundary, but lack of it does not. Think, for instance, of the set F defined by the fractions of *one* divided by any positive integer, that is, $F = x/x = \frac{1}{q}, q \in Z_+, q \neq 0$; as the value of q increases, the value of the corresponding element in the set decreases and it is not possible to find a smallest element; however, zero is a lower boundary. Actually the set will approach zero asymptotically but never reach it. This leads us to the statement that if a set has a smallest element, it is unique.

In the context of highway ramps, traffic flow will define a space; as you can take the difference between any two values, it also has a lower limit as the minimum observed flow is zero vehicles per hour; however, does it have an upper limit?

At first glance, you may think that a highway can accommodate any number of vehicles and therefore there is no upper limit. However, as you will learn from your transportation classes, yes there is; typically, levels of service can go up to a maximum flow of about 1900 veh/hour per lane (called level of service E) and then decline as there is no sufficient space to accommodate more vehicles, thereby resulting in traffic jams.

Other examples of similar systems are given by infrastructure elements. Think, for per instance, of the flow of water on a storm sewer pipe. Is there a minimum value of flow? Yes zero, no rain! Is there a maximum flow? Yes, when the pipe is at full capacity (that is why there is flooding). Now let us move into a more interesting problem: imagine a landfill (garbage disposal facility). They are comprised of a piece of land, which is restricted in surface by its boundary limits and by volume to the maximum elevation that the local regulations and the maximum side slopes will allow you to rise the landfill hill. Is there a maximum capacity for a landfill? Yes, it is given by its volume, and the rate of arrival of garbage will determine how many days the landfill has left. Can I push the capacity beyond this upper limit? Only if you manage to alter its very definition, that is, purchase more land or change the regulations that limit it.

2.1.5 Open Balls, Open Sets, Closed Sets

In characterizing the feasible space, we need to identify upper and lower boundaries and to characterize the very fabric of the space itself. As it will soon become evident, these characterizations of the space require consideration of two concepts: closeness and compactness.

The need for defining a closed space is tied to the definition of compactness and both to the existence of a solution for an optimization problem.

Let us define a useful element called a ball. A ball in two dimensions is simply a circle centred at x_1 with radius r_1, and an open ball is the one that does not include its boundary.

An open set is the one for which I can always find an open ball, no matter how close I get to its limits. In other words, there is always an element q, $q \neq x_1$ which is part of an open ball centred at any x_1 of the set, with some radius r_1. Notice that here the key is to take on small values of r_1 as one approaches the limit of the set.

For example, remember the case of $F = 1/q, q \in Z_{+}, q \neq 0$; the set will be open at zero. In general, the definition of open set corresponds to the one that does not include its boundary, and therefore, there will always be an element between x_1 and the limit of the set (zero for the example).

One way to know if a set is closed is to check if its complement is open. Let us take an example: let us use the real numbers between one and five [1,5]. Is this an open set?, to answer it is take an open-ball centered at one, and use any radius, you will see that such ball irremediably reach beyond the limit of the set. In other words, there does not exist an element between the first value of the set and its limit; that is, in simple words, one can 'walk' on the edge of the set's limit.

The problem with open sets is that one cannot reach their limit, and this creates an ill-defined problem because very often one desires to only consider solutions at the boundary of the feasible space as we will see in Chapter 3.

Hence, open intervals are open sets and closed intervals are closed sets. Another interesting definition is that of interior and boundary points, and more generally limit points. Let $A \subset X$ and $x \in X$; then x is a limit point of A if every neighbourhood (open ball) of x intersects with A at a point other than x. In other words, p is a limit point of a space A if $\exists\ q \in A$ such that $q \in B(p, r)$ for any $r > 0$. Start by moving a point x a bit ϵ inside the space A.

It is easy to see that if the space is continuous, the intersection of any open ball centred at x with radius smaller than ϵ will contain some points and help x satisfy the definition of a limit point. This kind of point is also an interior point.

If we return to the starting position and move x a bit ϵ away but this time outside the region A, we can see that for open balls centred at this new location of x and with radius smaller than ϵ the intersection of such open ball and the space A is empty.

So any interior point or boundary point is a limit point if A is closed. In practical terms, the definition of such points helps us identify the portions of the space we can work with (feasible space). But if A is open, the points at its boundary will not be elements of A; still they are limit points.

Notice the definition of a limit point does not require the limit point to be also part of the space. Hence, a closed space contains all its limit points. An open space will not. This simple definition will be very useful for quickly identifying the continuity and differentiability nature of some problems.

Let us look at some examples. Imagine we define the space for the amount of contracts a company may take to construct bridges in Canada. It turns out that such construction company can take on the construction of $1, 2, 3, \ldots, n$ bridges but not on the construction of 0.1238765392 of a bridge, which would be catastrophic as the company's reputation will indicate they barely did some work. When making decisions, this company will know that its decision variable is the number of bridges they will construct per year. Hence, the space A for bridge construction is given by integers.

Even though this observation appears trivial, it is not when solving for an optimization problem in which the levels of several decision variables must be chosen. Think of it as what makes a bridge: a superstructure, a substructure and a deck!, and in turn the deck requires materials (formwork, concrete, steel) in addition to the use of machinery and labour.

So the choice of constructing a bridge implies the use of other resources which are limited. Is the space of concrete continuous? It appears yes; we could select any amount of concrete measured in weight or volume. Is the space of labour continuous? Is it that of machinery?

Let us imagine now that we can break down everything in unitary terms and say that each linear meter of a 2-lane bridge requires 200 m^3 of concrete, 50 steel rebars type 40M, 100 m^2 of formwork, 10 hours of machinery work and 5 people for 20 hours. Can you tell if the space of labour is closed?

It turns out that it is preferable to work with closed spaces because one can take any value within the space. The problem of an open space is that values at the boundary cannot be used; it is like seeing the $100 bill on the floor and not being able to grab it.

An example of an open space would be that of the maximum number of revolutions on your car dial; it is there, you can see it, but I really do not advice you to go there. Luckily, most of the cases will involve closed spaces.

2.1.6 Compactness

There exist several definitions of compactness. For the purpose of this book, I will limit compactness to the most simple one. A space is compact if it is both closed and bounded. This is known as the Heine–Borel definition.

A compact space corresponds to a feasible space that is 'well behaved'. Hence, in order to have a well-behaved optimization problem, we require a compact space and a 'well-behaved' objective function.

Defining a decision variable (control variable) is an intrinsic requisite of having a compact space. Decision variables need to be characterized in terms of their membership to either the real, natural or integer numbers.

2.1.7 Convexity

A space is convex if for every pair of points within the object, every point on the straight-line segment that joins the pair of points is also within the space. For example, a solid cube is convex, but anything that is hollow or has a dent in it is not.

A function defined on an interval is called convex if the line connecting any two points on the graph of the function lies above the graph (plot) of the function. It is very important to quickly visualize convexity. The idea of convexity is particularly useful in unbounded problems because a solution may exist when the objective function is convex and the feasible space is concave.

Concave simply follows the contrary; any line connecting any two points on the graph of the function lies below the graph (plot) of the function.

2.1.8 Plotting Objects

As you can see, the ability to graph objects on space becomes very useful when dealing with open, bounded, convex and continuity concepts. The ability to plot any given function is particularly useful for quick visualization.

This section provides the reader with the steps required to plot on MATLAB®. I will look at two specific examples. The reader can possibly find more examples online. MATLAB contains three main windows: the history window, the command window and the workspace.

The workspace is commonly used for the definition of subroutines, we will not use it in this book. The command history keeps track of the commands you have used. The command window is where we input the commands I will cover in the following.

Let us take as first example plotting of the following function: $y = 2 + \log(x)$.

First, you need to define the range for any variable. On the *command window*, type $x = 0.5 : 0.01 : 3$; this will define the variable x starting at 0.5 ending at 3, with increments of 0.01.

Second, you need to define the function y, so type on the *command window* $y = 2 + \log(x)$.

Third, you can plot now, so type on the *command window plot(x,y)* and hit *enter*. You will obtain a graph.

Four, if you want to add a title for the graph, so type on the command window *title('Plot of y versus x'), xlabel('x'), ylabel('y')* and hit enter. You will obtain the graph shown in Figure 2.1.

To clear the command window, type *clc*; to clear the variables, type *clear*. Now let us plot several lines into the same graph; presume you have the following feasible space defined by the following set of equations:

$$x_1 + x_2 \geq 1,$$

$$x_1 \geq 0,\ x_2 \geq 0$$

and

$$x_1 + x_2 \leq 10$$

We define first the range for x_1 and x_2; as you can see, they go from zero to any number, but for practical purposes, we will define them up to a value of 11 given that there is a line crossing at an intercept of 10 with both axes. So we type on the command window $x_1 = 1 : 11$ and hit enter, then $x_2 = 1 : 11$ and hit enter.

The first equation needs to be expressed in terms of x_2, so it becomes $x_2 = 1 - x_1$; in MATLAB, we type $x_2 = 1 - x_1$. For the third equation, we

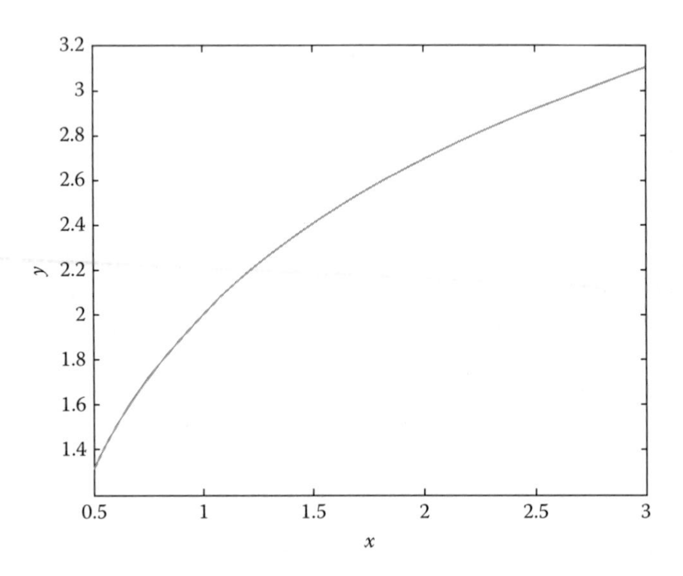

FIGURE 2.1
MATLAB® plot of y versus x.

need to create a new variable to store the relationship; we call it x_3 and type $x_3 = 10 - x_1$; as you can see, I have defined everything in terms of x_1. For the second equation, you already defined $x_1 = 1{:}10$; now you need to define the line that corresponds to zero, so you type a new variable, say x_4, and you type $x_4 = 0 * x_1$.

On MATLAB, we need to indicate in the software the need to plot everything on the same graph; for this purpose, we use the command *hold on*. Here are the MATLAB commands in the required sequence to produce the graph:

$$x_1 = 1 : 10$$

$$x_2 = 1 - x_1$$

$$x_3 = 10x_1$$

$$x_4 = 0 * x_1$$

hold on; plot$(x_1, x_2, \text{'}r\text{'})$; *plot*$(x_1, x_3, \text{'}b\text{'})$; *plot*$(x_1, x_4, \text{'}g\text{'})$; *hold off*.

After this, you produce the following graph. Notice that the code 'r' stands for red color line, 'b' stands for blue and 'g' stands for green. You can alternatively define labels and title using *title('Plot of Feasible space'), xlabel('*x_1*'), ylabel('*x_2*')*. Figure 2.2 shows the end result.

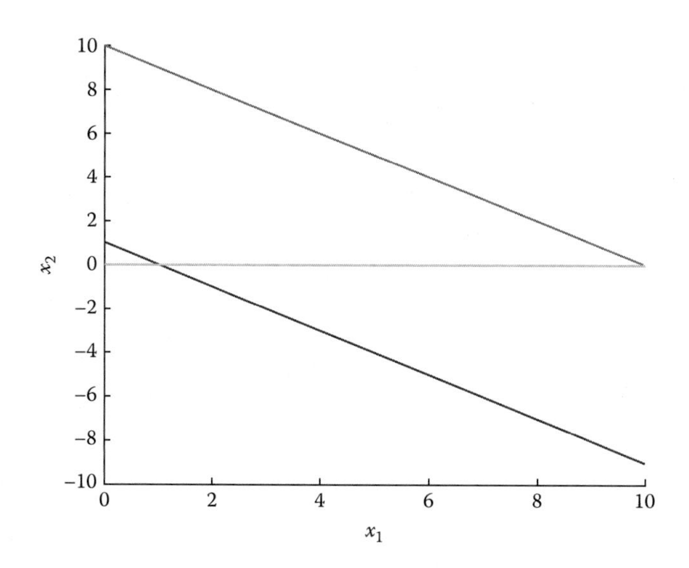

FIGURE 2.2
MATLAB plot of the feasible space.

2.2 Characterizing an Objective Function

A well-behaved objective function is continuous and differentiable. A function is continuous if for every open space in y the inverse image is an open space in x. This concept relies on the ability to have an inverse and for the purpose of this book can be simplified as the ability to map from any value in y to x. Inverse images of open sets are open. Picture a parabola and take any open space (i.e. on the y-axis, select an open interval). If the function is continuous, the inverse image will result in an open interval on the x-axis.

A function is continuous if for every convergent sequence with limit x the sequence converges to $f(x)$.

Differentiability refers to the ability to take derivatives. For most of the problems studied in this book, we require a function to be twice differentiable, the first time to have first-order conditions. If a function is differentiable, then it is continuous.

2.2.1 Warranting Optimality

If a feasible space is compact and an objective function is continuous, then by Weierstrass theorem, the function attains a maximum and minimum on such space. This theorem is one of the battle horses of optimization and is the motivation behind careful definition of both feasible space and objective function.

Even though second-order conditions exist, I will deliberately ignore them in this textbook.

2.3 First-Order Conditions

You probably remember from your senior high school years that in your calculus class you learn something regarding how to find a maximum or a minimum.

First-order conditions refer to those regions of a function where the slope reaches a value of zero. The technique used to identify the point where the slope is zero is by taking the derivative of the function and making it equal to zero. This is however applicable to two-dimensional spaces.

For higher-dimensional spaces, we proceed in a similar way: we can take the partial derivative with respect to each variable and equalize it to zero.

The first-order condition can be used to identify the circumstances under which a certain variable can be optimized. If you are interested in finding out whether you reached a maximum or a minimum, then you may need to use the second derivative.

2.4 Lagrange Method

The Lagrange method is amply used in many fields. The method combines the objective with the constraints into one equation and uses first-order conditions to solve. Typically out of each first-order condition, you obtain one expression (equation) that gives you a representation of an optimal behaviour.

The Lagrange method can be summarized in the following steps:

1. Write down your objective.
2. Every constraint is multiplied by a parameter, the typical name for the parameters is lambda λ. If you have several constraints, then use one λ per constraint (λ_i). Sometimes, the time dimension may be present, then call the parameter λ_i^t where i stands for the constraint and t for the time period. In the most general case, you could have one $\lambda_{i,j}^t$, where i is the decision variable, j is the individual for which we are solving and t is the period of time. We would not look

at this level of detail; you could encounter this at recursive dynamic macroeconomic problems.

3. Apply first-order condition; that is, take a partial derivative with respect to each decision variable and equal each to zero.
4. Solve and obtain an expression for each decision variable.

Each optimization problem needs to be formulated as you saw in the previous chapter; the same is applicable here. Learning the mechanisms behind the nature of a problem is never easy; the more problems you formulate, the easier it will be to identify the bare bones of new problems. I will now introduce two examples—the formulation will be completely explained in the first one while in the second example the formulation will be given.

The first example is a very basic one. Consider a water tank of cylindrical dimensions given by a radius r for the bottom and a height h for the wall. The wall of the tank can be seen as a rectangle of dimensions $2\pi r$. For the top and the bottom of the tank, you need circles. Figure 2.3 illustrates the initial set-up for the metal sheet cutting.

We will think of them as squares to simplify the process. The problem is how to maximize the volume of the tank while using a square sheet of metal to assemble it. Figure 2.4 illustrates the abstraction of the square metal for this problem. We start by placing two lids and a wall over the metal sheet.

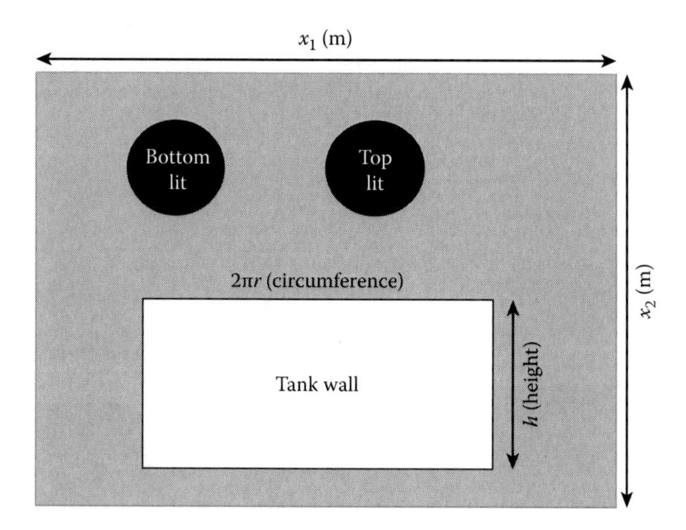

FIGURE 2.3
Initial set for metal sheet.

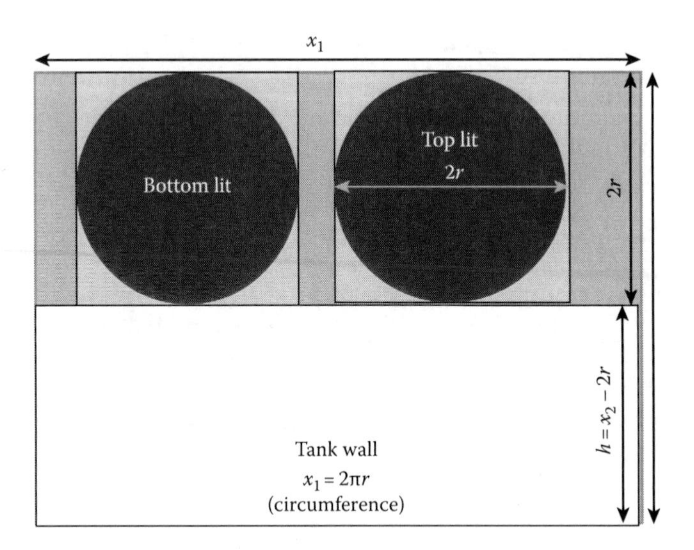

FIGURE 2.4
Assuming squares for top and bottom.

The next thing you probably notice is that the squares needed for the bottom and top are of dimensions $2r$. I have place them one next to the other one.

As you can see, I have written the height of the tank as a function of the radius and the given side size as x_2. We now try to formulate the objective: to maximize the volume. The volume is given by the tank height $x_2 - 2r$ multiplied by the area of the bottom (and top, they are the same) given by π multiplied by r^2. The only decision variable in this optimization is the radius, so we want to find its value. The maximization takes the following form:

$$(x_2 - 2r)(\pi r^2).$$

The constraints are shown in Figure 2.5; as you can see, we have one constraint per dimension of the metal sheet. For x_2, we know that it will be equal to $h + 2r$; however, we notice we do not know what h is, so it becomes a second decision variable. For x_1, we know that it has to be smaller or at most equal to $2r + 2r = 4r$.

The final formulation is as shown in the following. We have one objective and two constraints.

$$\max_{r}(x_2 - 2r)(\pi r^2).$$

$$x_1 - 2\pi r = 0 \quad x_2 - 2r - h = 0.$$

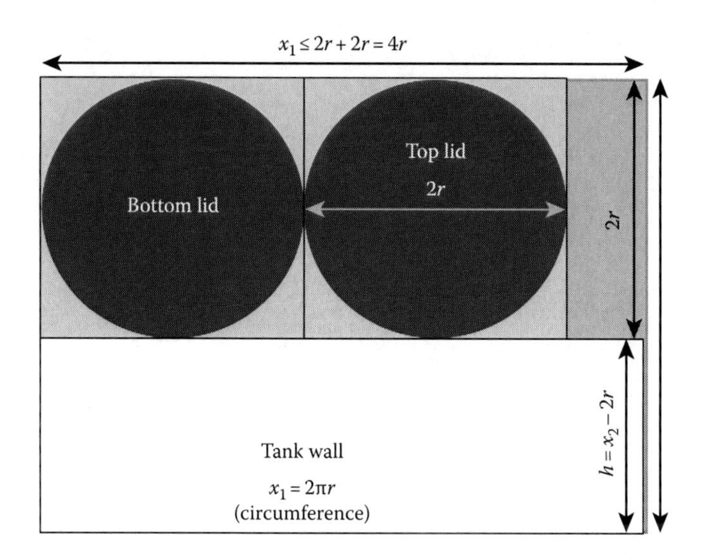

FIGURE 2.5
Constraints on metal sheet.

We use the Lagrange method to solve. First, we create the Lagrangian, an equation that puts together both the objective and the constraints.

$$L = \max_r(h)(\pi r^2) + \lambda_1(x_1 - 2\pi r) + \lambda_2(x_2 - 2r - h).$$

Our objective is to estimate the values of the radius r and the height h in terms of x_1 and x_2. The next step in the Lagrange method (step 3) is to take partial derivatives with respect to each decision variable and with respect to each Lagrange multiplier and equal to zero. This corresponds to the following first-order conditions:

$$\frac{dL}{dr} = 2\pi hr - 4\lambda_1 - 2\lambda_2 = 0,$$

$$\frac{dL}{dh} = \pi r^2 - \lambda_2 = 0,$$

$$\frac{dL}{d\lambda_1} = x_1 - 2\pi r = 0,$$

$$\frac{dL}{d\lambda_2} = x_2 - 2r - h = 0.$$

The last step (step 4) is to solve the system of equations and find the values of the decision variables (r and h in this case). We take the third equation and directly obtain the value for $r = \frac{x_1}{2\pi}$. We input this explicit value of r into the last equation and solve for h:

$$x_1 - 2\pi r = 0$$
$$r = \frac{x_1}{2\pi}$$
$$x_2 - 2r - h = 0$$
$$x_2 - 2\left(\frac{x_1}{2\pi}\right) = h$$
$$x_2 - \frac{x_1}{\pi} = h$$

The amounts x_1 and x_2 are given to you; they are the dimensions of the metal sheet. Therefore, the radius r of the bottom and top depends on x_1 and the height depends on both x_1 and x_2.

2.5 Gradient Search

The gradient search is a technique to find an optimal point. This technique follows an iterative process that changes the value of the variables until you are close enough to a maximum or a minimum given the satisfaction of the first-order condition.

This formulation of this technique can be explained using the Lagrange method.

The idea behind the gradient search is to find the optimal step size s to walk when trying to approach an optimal point. The discrete steps Δx and Δy taken in the x and y directions lead to a total change in the objective Δz which is approximately given by

$$\Delta z = \frac{\delta z}{\delta x}\Delta x + \frac{\delta z}{\delta y}\Delta y.$$

The partial derivative is assessed at the current location c for both directions (x and y). The objective is constrained by Pythagoras as follows:

$$(\Delta x)^2 + (\Delta y)^2 = (s)^2.$$

This can be formulated using the Lagrange method, and our decision variables will be Δx and Δy. The Lagrangian is

$$L = \frac{\delta z}{\delta x}\Delta x + \frac{\delta z}{\delta y}\Delta y + \lambda(s^2 - \Delta x^2 - \Delta y^2).$$

Taking partial derivatives, we get the following:

$$\frac{\delta L}{\delta(\Delta x)} = \frac{\delta z}{\delta x} - 2\lambda\Delta x = 0,$$
$$\frac{\delta L}{\delta(\Delta y)} = \frac{\delta z}{\delta y} - 2\lambda\Delta y = 0,$$
$$\frac{\delta L}{\delta\lambda} = (s^2 - \Delta x^2 - \Delta y^2) = 0.$$

Solving for λ in the first two equations, we get the following:

$$\lambda = \frac{1}{2\Delta x}\frac{\delta z}{\delta x},$$
$$\lambda = \frac{1}{2\Delta y}\frac{\delta z}{\delta y}.$$

Equal and solve

$$\frac{1}{2\Delta x}\frac{\delta z}{\delta x} = \frac{1}{2\Delta y}\frac{\delta z}{\delta y},$$
$$\frac{\delta z/\delta x}{\delta z/\delta y}\Delta y = \Delta x.$$

Substituting into the step size equation given by Pythagoras, we find

$$\frac{s\dfrac{\delta z}{\delta y}}{\sqrt{\dfrac{\delta z}{\delta y}^2 + \dfrac{\delta z}{\delta x}^2}} = \Delta y.$$

Similarly,

$$\frac{s\dfrac{\delta z}{\delta x}}{\sqrt{\dfrac{\delta z}{\delta y}^2 + \dfrac{\delta z}{\delta x}^2}} = \Delta x.$$

These two equations are written in terms of a step size s and the values of slopes at specific points; all values are known and the only unknown is the step size. You can take and test this value and monitor the value of the derivative until it approaches zero. Each time you move, you need to update the current value of $x_{i+1} = x_i + \Delta x$ and of $y_{i+1} = y_i + \Delta y$. Let us look at an example; consider the following equation: $Z = x^2 + y^2 - 4$. This equation can be graphically represented by Figure 2.6.

The gradient search method can be set up in Excel in order to estimate both the value of Δx and Δy and the corresponding value of the derivative (which hopefully will approach zero). Figure 2.7 shows a screenshot of the main interface created in Excel and the codes used to achieve both purposes mentioned previously.

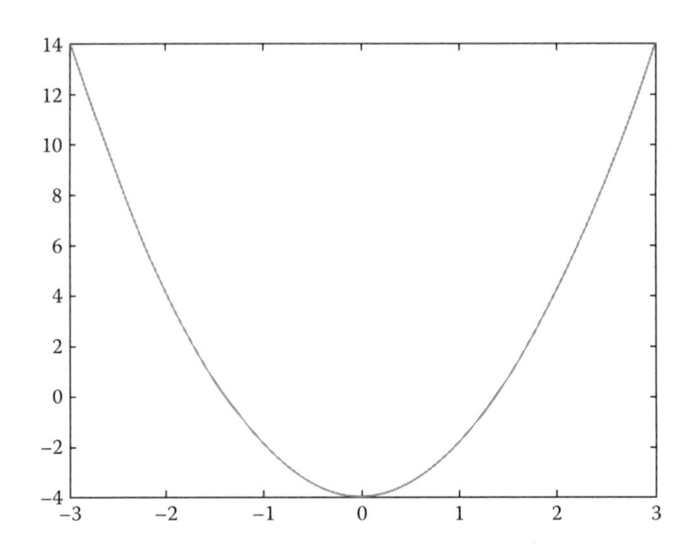

FIGURE 2.6
Gradient example: graphical representation.

SUM | × ✓ f_x | =+B2+(B5*(2*B2)/(SQRT(4*B2^2+4*B3^2)))

	A	B	C	D	E
1		Initial = 0	1	2	3
2	x	1	=+B2+(B5*(2*B2)/(SQRT(4*B2^2+4*B3^2)))	0.01005	0.002979
3	y	1	0.292893219	0.01005	0.002979
5	s	-1	s2= -0.4	-0.01	
6					
7	x'		0.585786438 = 2 c2	0.0201	0.005959
8	y'		0.585786438 = 2 c3	0.0201	0.005959

FIGURE 2.7
Excel spreadsheet for gradient search.

If we take partial derivative, it is easy to see that $\frac{\delta z}{\delta x} = 2x$, which in Excel translates to 2C2 given that C2 is the corresponding cell with the current value for x.

Similarly for y, we obtain that the partial derivative is $\frac{\delta z}{\delta y} = 2y$, which in Excel translates to 2C3 given that C3 is the corresponding cell with the current value for x.

The explicit forms for the Δx and Δy increments considering the previous derivative are the following:

$$\frac{B5 * (2 * B2)}{\sqrt{4 * B2^2 + 4 * B3^2}} = \Delta y$$

and

$$\frac{B5 * (2 * B3)}{\sqrt{4 * B2^2 + 4 * B3^2}} = \Delta x.$$

These equations are only measuring the differences, so they have to be added to the initial amounts at each step. This is achieved by simply adding at the beginning the initial amount as shown before and as shown by the Excel spreadsheet in Figure 2.7:

$$x + \frac{B5 * (2 * B2)}{\sqrt{4 * B2^2 + 4 * B3^2}} = \Delta x$$

and

$$y + \frac{B5 * (2 * B3)}{\sqrt{4 * B2^2 + 4 * B3^2}} = \Delta y.$$

Exercises

1. Define the control variables in the following cases and identify the feasible space boundaries.
   - **a.** Pipe to conduct rainwater
   - **b.** Traffic flow of one lane at a given bridge
   - **c.** Bearing soil capacity for the design of a foundation
   - **d.** Concrete strength for the design of a column
   - **e.** Water pressure on a system of water mains
2. Characterize the objective function in the same circumstances as before with the following additional information.

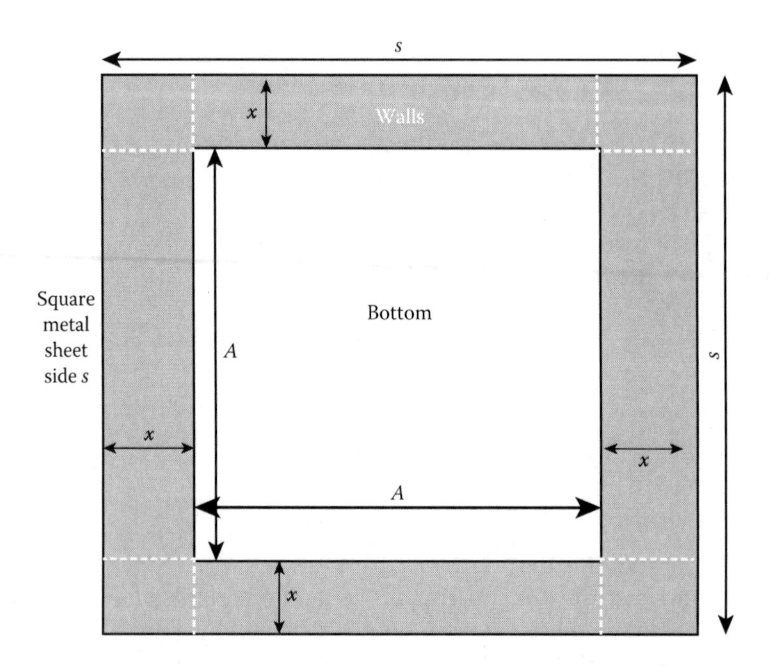

FIGURE 2.8
Metal sheet for Exercise 3.

3. Solve using the first-order condition (FOC) method. A parallelepiped water tank (think of a square-like section or a box) made out of a square metal sheet of side s as shown in Figure 2.8. The box has no lid (no top cover). Define the objective and constraints and solve using the FOC method.
4. Solve the following using the Lagrange method. Consider a parallelepiped water tank (think of a rectangular-like section) made out of a rectangular metal sheet as shown in Figure 2.9. The box has no lid (no top cover). Define the objective and constraints and obtain the derivatives of step 3 using the Lagrange method. Try to solve (hint: replacing the constraints into the objective gives you a fast-track way to solve).
5. For the previous example, obtain an expression for the optimal area size for the bottom of the box. Formulate the objective (and constraints if necessary) and solve to find the optimal value of C to cut the metal sheet. Use this additional information to solve Problem 4.
6. Formulate and solve the following problem.

   A driver on a highway is using gravity; maximum speed on the highway is 100 km/h and, minimum speed is 60 km/h. The driver accelerates to about 140 km/h; then shifts from 6 to neutral (manual gear car) and lets the car roll/go downhill by gravity until friction slows down his car to about 80 km/h. By doing this, the car engine's revolutions (cycles

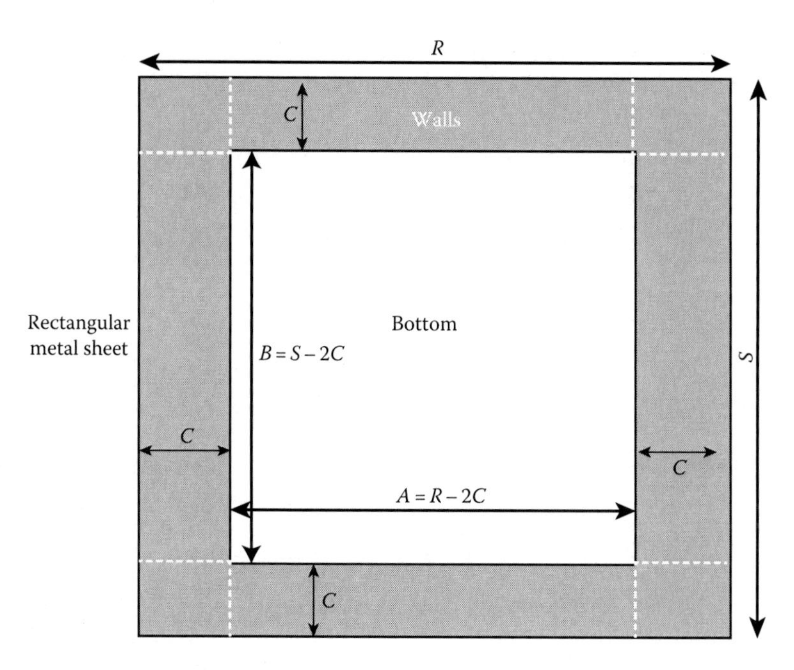

FIGURE 2.9
Metal sheet for Exercise 4.

per second) drop from about 2.0 rpm to about 1.0 rpm, so fuel consumption drops by 50% during the time the car is in neutral.

Assuming you are optimizing the driver behaviour, what is the minimum amount of time $T1$ the driver should go at 1.0 rpm in order to optimize saving. Assume \$1 per minute is the average fuel saving when the vehicle goes at 1.0 rpm; let $T1$ be the amount of time the car goes at 1.0 rpm. Suppose the likelihood of getting a ticket of \$220 is 1%, the likelihood of crashing for any reason is 0.1% and the mean estimated cost of crash is \$2500. Let $(T1/2)^2$ be the amount of time the driver is going at a speed higher than the allowed, and therefore being exposed to a ticket or lacking proper control to avoid a collision.

Formulate the objective as maximization of savings, and formulate any constraints (if applicable) and solve using first-order condition.

7. Identify the feasible space (defined by the constraints) and provide an explanation if the space is closed or open, and bounded:
   a. The feasible space is defined according to compliance with the following set of constraints: $x_1 + x_2 \geq 1$, $x_1 \geq 0$, $x_2 \geq 0$ and $x_1 + x_2 \leq 10$. Hint: Plot the constraints.
   b. The feasible space is defined according to compliance with the following set of constraints: $x_1 + x_2 \geq 1$, $x_1 > 0$, $x_2 \geq 0$ and $x_1 + x_2 \leq 10$. Hint: Plot the constraints.

   c. The feasible space is defined according to compliance with the following set of constraints: $x_1 + x_2 > 1$, $x_1 \geq 0$, $x_2 \geq 0$ and $x_1 + x_2 \leq 10$. Hint: Plot the constraints.

   d. The feasible space is defined according to compliance with the following set of constraints: $x_1 + x_2 > 1$, $x_1 \geq 0$, $x_2 \geq 0$. Hint: Plot the constraints.

   e. The feasible space is defined according to compliance with the following set of constraints: $x_1 + x_2 > 1$, $x_1 \geq 0$, $x_2 \geq 0$ and $x_1 + x_2 \geq 10$. Hint: Plot the constraints.

   f. The feasible space is defined according to compliance with the following set of constraints: $x_1^2 + x_2^2 \leq 1$, $x_1 \geq 0$, $x_2 \geq 0$. Hint: Plot the constraints.

   g. The feasible space is defined according to compliance with the following set of constraints: $x_1^2 + x_2^2 \geq 1$, $x_1 \geq 0$, $x_2 \geq 0$ and $x_1 + x_2 \leq 10$. Hint: Plot the constraints.

   h. The feasible space is defined according to compliance with the following set of constraints: $x_1^2 + x_2^2 > 1$, $x_1 \geq 0$, $x_2 \geq 0$ and $x_1 + x_2 \leq 10$. Hint: Plot the constraints.

   i. The feasible space is defined according to compliance with the following set of constraints: $x_1^2 + x_2^2 \geq 1$, $x_1 > 0$ and $x_1 + x_2 \leq 10$. Hint: Plot the constraints.

8. Identify in each of the previous cases where do you have a compact feasible space.

9. The following are objective functions for each of the items under Problem 7. Identify if the objective is continuous, and then identify by Weierstrass theorem if there must be an optimal point (either minimum or maximum).

   a. $x_1 + 2x_2$.

   b. x_1 if $x_1 < 2$ and x_1^2 if $x_1 \geq 2$.

   c. $x_1 + 3x_2$ if $x_2 < 1$ and $3x_1 + x_2$ if $x_2 \geq 1$.

   d. $x_1 + 3x_2$ if $x_2 < 1$ and $3x_1 + x_2$ if $x_2 \geq 2$.

   e. $2x_1 + 2x_2$.

   f. $x_1^2 + x_2^2$.

10. Identify if the space is convex:

   a. The feasible space is defined according to compliance with the following set of constraints: $x_1 + x_2 \geq 1$, $x_1 \geq 0$, $x_2 \geq 0$, $x_1 + x_2 \leq 10$. In addition, the space comprised between the following two lines must be removed from the feasible space $x_1 + x_2 > 3$ and $x_1 + x_2 < 4$.

   b. The feasible space is defined according to compliance with the following set of constraints: $x_1 + x_2 \leq 1$, $x_1 \leq 0$, $x_2 \leq 0$ and $x_1 + x_2 \geq 10$. Hint: Plot the constraints.

c. The feasible space is defined according to compliance with the following set of constraints: $x_1 + x_2 \geq 1$, $x_1 \geq 0$, $x_2 \geq 0$ and $x_1 + x_2 \leq 10$. Hint: Plot the constraints.

d. The feasible space is defined according to compliance with the following set of constraints: $x_1^2 + x_2^2 \leq 1$, $x_1 \geq 0$, $x_2 \geq 0$. Hint: Plot the constraints.

e. The feasible space is defined according to compliance with the following set of constraints: $x_1^2 + x_2^2 \leq 4$ and $x_1^2 + x_2^2 \geq 1$, $x_1 \geq 0$, $x_2 \geq 0$, $x_1 + x_2 \geq 10$ and $x_1 + x_2 \leq 12$. Hint: Plot the constraints.

Solutions

1. a. Decision variable = flow capacity on the pipe, space defined on positive real numbers. Lower boundary = zero, upper boundary = largest market-available pipe size.

   b. Decision variable = number of vehicles per hour per lane, space defined on positive integer numbers. Lower boundary = zero, upper boundary = commonly 1900 veh/hour/lane.

   c. Decision variable = foundation dimensions in centimetres, space defined on positive integer numbers (notice you would not design half a centimetre, and as a matter of fact, your design will be conditioned by rebar length in discrete jumps of dimensions). Lower boundary = zero, upper boundary = largest possible foundation size as restricted by neighbour properties or foundation budget.

   d. Decision variable = concrete strength, space defined on positive real numbers. Lower boundary = zero, upper boundary = largest market-available concrete strength. Notice in this case that concrete strength could take any value with decimals although the prescription when you buy goes on finite intervals.

2. Objective is defined as

   a. The objective is to maximize pipe capacity.

   b. The objective is to maximize traffic flow (number of vehicles per hour).

   c. The objective is to maximize bearing capacity.

   d. The objective is to maximize concrete strength.

   e. The objective is to maximize water pressure.

3. Figure 2.10 shows the relationships between the side size and the unknown amounts for the bottom side (A) and wall height (x).

   The volume is $\max_{x}(S - 2x)(x)$.

   The partial derivative takes the form

$$\frac{dL}{dx} = S - 4x = 0.$$

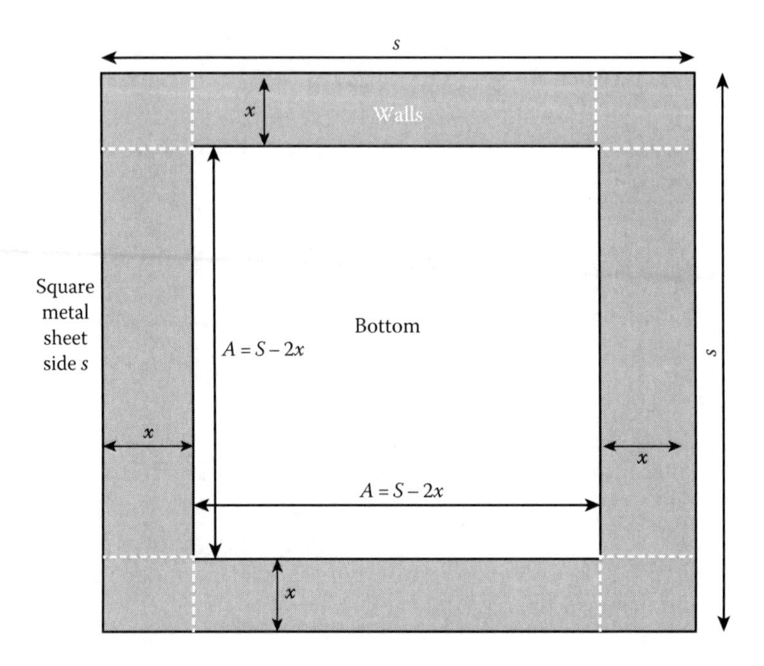

FIGURE 2.10
Solution of Exercise 3.

From this, we see that x takes the value of $\frac{S}{4}$, and this will return in the optimal volume (maximum) for the water tank.

4. In this case, you probably notice that x must be equal to y; in other cases, you would have an uneven wall height, so I will only use only x from now on. The volume is

$$\max_{ABC} ABC.$$

Subject to the following constraints:

$$S - 2C - B = 0,$$
$$R - 2C - A = 0.$$

We create the Lagrangian:

$$L = \max_{A,B,C} ABC + \lambda_1(S - 2C - B) + \lambda_2(R - 2C - A).$$

Take partial derivatives with respect to each unknown (i.e. A, B, C) and the Lagrange multipliers:

$$\frac{dL}{dA} = BC - \lambda_2 = 0,$$
$$\frac{dL}{dB} = AC - \lambda_1 = 0,$$
$$\frac{dL}{dC} = AB - 2\lambda_1 - 2\lambda_2 = 0,$$
$$\frac{dL}{d\lambda_1} = S - 2C - B = 0,$$
$$\frac{dL}{d\lambda_2} = R - 2C - A = 0.$$

Remember that both R and S are given, which are the dimensions of the metal sheet used to create the water tank, we are interested in finding the values of $A, B, C,$ not the value of the Lagrange multipliers. Unfortunately, we need to find these values to be able to solve.

$$BC - \lambda_2 = 0 \text{ then } BC = \lambda_2,$$
$$AC - \lambda_1 = 0 \text{ then } AC = \lambda_1.$$

We take the first equation and replace the Lagrange multipliers.

$$AB - 2AC - 2BC = 0,$$
$$\frac{AB}{2} = C(A + B).$$

The last two equations tell us that if $2c = 2c$, then

$$S - B = 2C$$

and

$$R - A = 2C.$$

Therefore,

$$S - B = R - A \text{ and } S - R = A + B.$$

Given the fact that both S and R are known amounts, its difference is also a fixed known amount (such as 2, 7, 15, just to mention a few possibilities). So we end up with two equations that characterize the problem.

The first one tells us the relationship between the metal sheet and the dimensions of the bottom of the box:

$$B = \frac{S - R}{A}$$

and correspondingly,

$$A = \frac{S - R}{B}.$$

The second one establishes the size of the wall in terms of the dimensions:

$$C = \frac{AB}{2(A + B)}.$$

5. Figure 2.11 illustrates the solution for Exercise 5. The objective must consider two constraints: the first constraint is for the side A and is equal to $R = A + 2C$, the second constraint is for the other side B and is equal to $S = B + 2C$. We input this constraint directly into the equation for

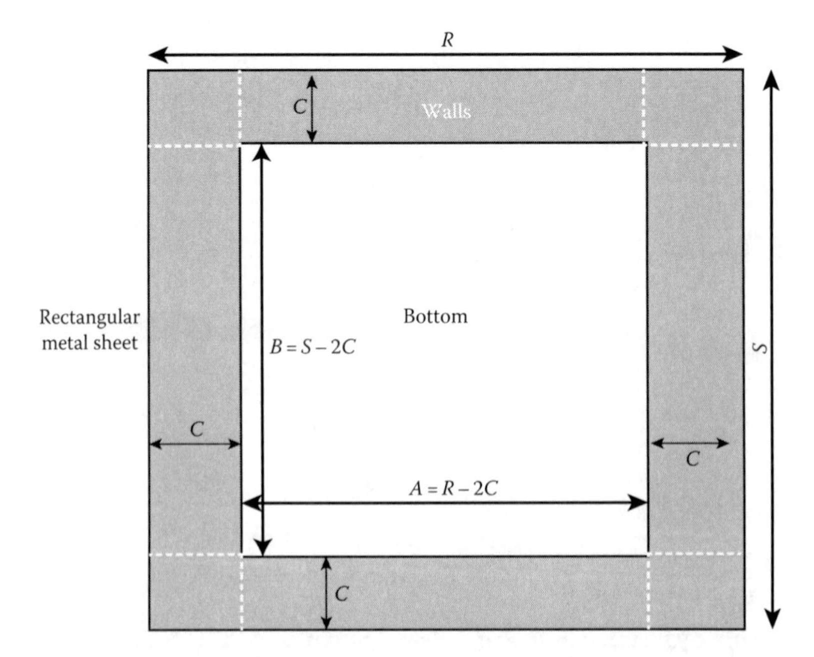

FIGURE 2.11
Solution of Exercise 5.

maximum area, which simply multiply both dimensions. We use the FOC to solve.

$$L = \max_C (R - 2C)(S - 2C),$$

$$L = \max_C RS - 2CR - 2CS + 4C^2.$$

This simpler form allows us to handle one decision variable C and solve for it.

$$\frac{dL}{dC} = -2R - 2S + 8C = 0.$$

We find that $-2R - 2S + 8C = 0$ leads us to $(R + S)/4 = C$. Both values R and S are known, so C is known. This result can be used back in Question 4 to find an expression for the optimal volume.

Recall that

$$S - B = 2C$$

and

$$R - A = 2C.$$

Knowing that

$$C = \frac{R + S}{4},$$

then

$$S - B = \frac{R + S}{2}$$

which means that

$$S - \frac{R + S}{2} = B.$$

Similarly,

$$R - \frac{R + S}{2} = A.$$

6. The cost can be expressed as (220 * 0.01 + 2500 * 0.001). This cost is applicable only when the driver is exceeding the maximum allowable speed.

That fraction of the total time is given by $\frac{T1}{2}^2$. The objective would then be

$$\max_{T1} T1 - \frac{T1}{2}^2 (220 * 0.01 + 2500 * 0.001).$$

Take a partial derivative with respect to $t1$ and equal to zero.

$$1 - T1(4.7) = 0 \text{ which results in } T1 = 1/4.7 = 0.2127.$$

This means that the optimal amount of time to exceed the speed is 21.27% of the time.

7. It is open when the inequality does not include the equal: that is, it uses either smaller or bigger, and the side in consideration is part of the boundary of the feasible space.

   It is unbounded when it misses one portion of the boundary. The following are the answers for each subitem of Question 6.

   a. Closed and bounded.
   b. Open and bounded.
   c. Open and bounded.
   d. Open and unbounded.
   e. Closed and bounded. Is one quarter of a circle on quadrant I.
   f. Closed and bounded. Is one quarter of a circle as bottom and line as top.
   g. Open and bounded. Is one quarter of a circle as bottom and line as top.
   h. Open and unbounded.
   i. Closed and bounded.

8. The answer for Question 8 is
   a. Compact
   b. Noncompact
   c. Noncompact
   d. Noncompact
   e. Compact
   f. Compact
   g. Noncompact
   h. Noncompact
   i. Compact

9. Objectives defined by multiple segments could imply discontinuities, however, in some cases the function may still be continuous. Let us look at the answers item by item.

a. Continuous, and yes Weierstrass warrants optimality.

b. Discontinuous.

c. Continuous, with change of slope; yes Weierstrass warrants optimality.

d. Discontinuous with missing segment.

e. Continuous; you have a straight line; yes Weierstrass warrants optimality.

f. Continuous; you have a circle again; yes Weierstrass warrants optimality.

10. Anywhere we have a hole or missing portion, there would be problems with convexity. Let us look at the case-by-case answers:

a. It is not convex; there is a hole (rectangular wedge missing).

b. It is not convex; the constraints actually define the hole inside and the feasible space outside.

c. It is convex.

d. It is convex.

e. It is not convex; there are two well-identified portions of the feasible space: one goes between two portions of a quarter of a circle and the other between two straight lines.

3

Optimization and Decision Making

3.1 Introduction

This chapter is devoted to optimization and techniques to support decision making. It explains two main methods: the graphical approach and the simplex algorithm. The simplex subdivides on two main methods: an equation-based approach and a simplex tableau. The simplex matrix-based approach is not included in this review, and interested readers can find it from other sources. The simplex algorithm is applicable to linear problems, and it is a method that departs from a point and walks on the feasible space until it finds the optimal solution, that is, a point that returns the highest (or smallest) value for a given objective. For the following problems, we will concentrate our attention to only one objective and several constraints. The second portion of this chapter is dedicated to explain some techniques to guide decision making.

3.2 Graphical Approach

The graphical approach is a simple method used to solve linear programming problems. This approach is limited to problems in two dimensions, which involves a maximum of two decision variables. In addition, we will use it only for the resolution of problems that are linear in nature, and a simple way to define a linear problem is through a combination of decision variables accompanied by constant parameters. For example, imagine that you are selling sandwiches ($3) and hot coco ($2) and you are trying to maximize income. Let us call the income with the letter y, the number of sandwiches with x_1 and the number of glasses of hot coco with x_2. Income will evidently come from selling your products, and then we can write an expression for income: $y = 3x_1 + 2x_2$. This expression is linear because variables x_1 and x_2 come alone; if they were 'accompanied' by a power or some function (say logarithm or exponential), then they would not be linear.

Let us look at a simple example. Suppose I decide to use my spare time after work to produce ceramic tiles and paving stones. The raw material I use is clay and water, water is free at my apartment (included in the lease) and I purchase clay from a local producer, but given my apartment storage, capacity I cannot hold on more than 22 kg of clay per week. Let us assume for each tile I need 2 kg of clay and for each paving stone I use 3 kg of clay. After these are produced, paving stones can be placed to dry anywhere on my apartment, meaning I do not have a storage capacity, but paving stones need constant water and a curing process that limits them to my bathtub, so I cannot dry more than 6 per week. To produce them, each ceramic tile takes 15 minutes and each paving stone takes 1 hour. I only have 6 hours per week for this messy enterprise. Why do I bother producing them? Well each tile gives me a revenue of $40 and each paving stone $50 after deducting all cost of production. My problem is that I do not know how much are the optimal amounts to produce.

This simple problem will be used repeatedly to illustrate several methods. In this section, we use it to show how the graphical part works.

First, identify the objective and the constraints. At this stage, I trust you can visualize that the constraints come from limitations and that those things that typically limit us have an upper bound associated to them. For instance, raw material to 22 kg/week, labour to 6 hours per week and storage of paving stones to 6 units/week. The objective is our wish to either maximize or minimize these constraints. In this case, our objective is profit, revenue, earnings and money into my pocket! So this depends on how many paving stones (say x_1) and ceramic tiles (say x_2) I produce per week. And as a matter of fact, I know each paving stone gives me $50, so the number of paving stones I finally decided to produce will give me $50x_1$ and the number of ceramic tiles $40x_2$ which added together define my profit objective.

As you can see, I defined my decision variables as x_1 and x_2, so the constraints will be given as in Table 3.1. It is advisable to select x_i with $i = 1, 2, 3, 4, 5, 6, \ldots$ as decision variable so that you can handle problems with many variables instead of using letters to represent variables and run out of alphabet at some point.

The storage constraint is self-explanatory, the feasible space has a 'wall' boundary at $x = 6$ and any lesser value is acceptable (including negative numbers). The best way to draw a constraint is by finding the value of the intercepts with the x and y axes and then drawing a line to join them.

TABLE 3.1

Constraints

Nature of Constraint	Constraint
Materials	$3x_1 + 2x_2 \leq 22$
Labour	$\frac{1}{4}x_1 + x_2 \leq 6$
Storage	$x_1 \leq 6$
Non-negativity	$x_1, x_2 \geq 0$

Intercepts are found when one of the variables is sent to zero. So, for instance, for the labour constraint, when $x_1 = 0$, $x_2 = 6$ and when $x_2 = 0$, $x_1 = 24$. For the materials constraint, when $x_1 = 0$, $x_2 = 11$, and when $x_2 = 0$, $x_1 = \frac{22}{3} = 7.33$.

Figure 3.1 shows the initial feasible space, considering only non-negativity constraints and its evolution as more constraints are applied.

We are only missing to bring the objective function into the picture. As found before, the objective consists of $50x_1 + 40x_2$. One easy way to draw this is by making it equal to any number and finding the slopes. A convenient one is 200 (the product of the coefficients of x_1 times x_2), so let us draw $50x_1 + 40x_2 = 200$. The method is just as with the constraints, so we will get intercepts at $x_1 = 4$ and $x_2 = 5$. There is, however, a caveat; we need to identify the sense of improvement for this line; this is given by the sign of the slopes. So in this case, both coefficients are positive, meaning increase on x_1 and increase on x_2 is beneficial. Figure 3.2 shows other senses of improvement for sample objectives.

To find a solution, we proceed to move the objective in the direction of improvement until we touch the very last piece of the feasible space, and just before leaving, when we are intersecting to the last point (or set of points), we will find the answer to the optimization problem. Figure 3.3 illustrates the advancement of the objective function through the feasible space until the solution to the optimization problem is found.

By looking at Figure 3.3, it is clear that the solution was found at the intersection of two constraints, the labour and materials constraint. These constraints are called binding constraints because we will use all the resources available from them, that is, we will use exactly 22 units of material and 6 units of labour. In the language of optimization problem, we will satisfy them with an equality. So we could have simply equalized both constraint equations and solved for both unknowns.

In this case, visualizing the answer is almost trivial and hence we could have even projected the optimal point back to the axis to find the solution to the problem. These two easy and familiar approaches to solve are only possible when dealing with two decision variables. I ask now: Can you tell me which are the binding constraints for the following problem? Or even more, can you plot this and find a solution to it?

$$5x_1 + 3x_2 - 7x_3 + \frac{1}{4}x_4 \leq 15$$

$$2x_1 + 1x_2 - x_3 + x_4 \geq 1$$

$$-25x_1 + 3x_2 + 7x_3 - \frac{28}{3}x_4 \leq 100$$

$$x_3 + x_4 \leq 200$$

$$5x_1 + 3x_2 \leq 150$$

$$x_1 + x_3 + x_4 \geq 15$$

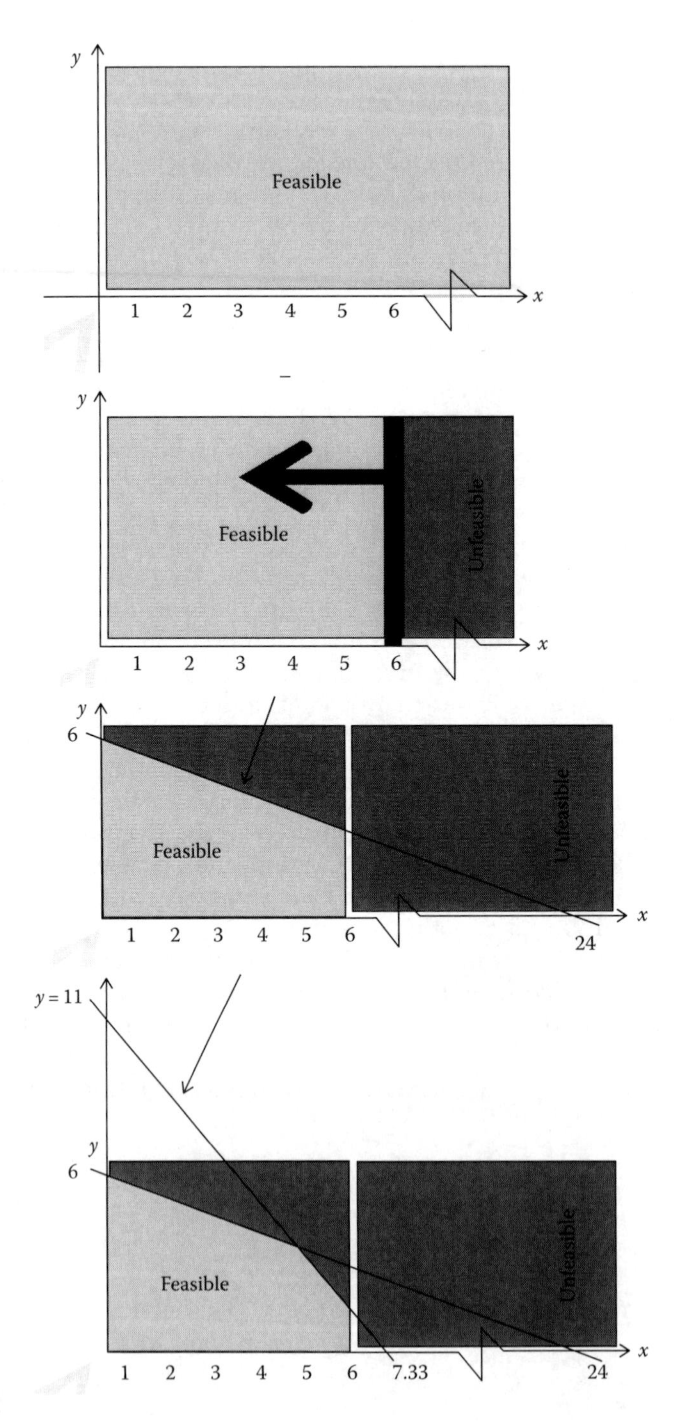

FIGURE 3.1
Feasible space.

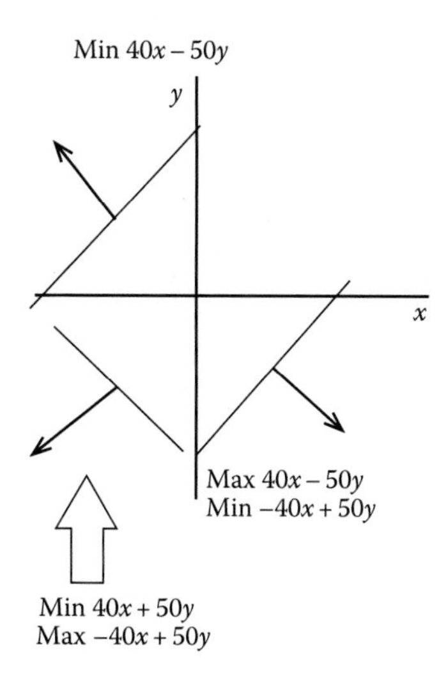

FIGURE 3.2
Moving the objective.

The very reason why we are learning a simplex method is to be able to solve these problems without the need to use the graphical approach or identify binding constraints.

3.3 Simplex Method

3.3.1 Setting Up the Problem

Before we get started, it is important to express all constraints as equalities; therefore, any inequality will be turned into an equality by adding a dummy amount called slack variable. This variable carries the difference between the limiting amount of resources and the actual amount being used (as a result of the optimization). While a constraint limited above by a certain amount will add a slack variable, a constraint limited from below will subtract the slack variable to force equilibrium between resource usage and its limit. Equations 3.1 and 3.2 illustrate both cases.

Consider what happens if $x_1 = 2$ and $x_2 = 3$, and then Equation 3.1 will result in $12 \leq 20$ which is true (holds); therefore, we could imagine that the difference $(20 - 12 = 8)$ is stored in a slack variable called S_1. Similarly, if we verify the same for Equation 3.2 (with the same values: $x_1 = 2$ and $x_2 = 3$), we will find that such equation arrives at a value of 21 and that $21 \geq 14$. Hence, it holds and we could capture the excess of seven units by a surplus variable called S_2 with the only difference that such variable will be subtracted in Equation 3.2.

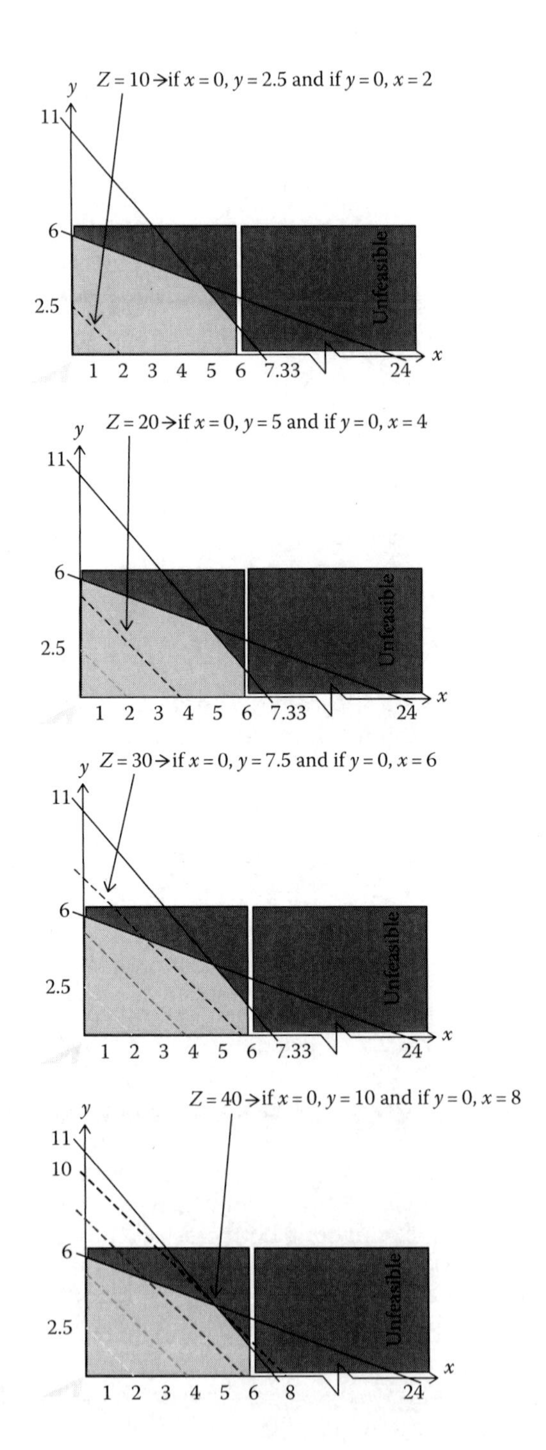

FIGURE 3.3
Finding the optimal point.

$$3x_1 + 2x_2 \leq 20 \text{ would turn into } 3x_1 + 2x_2 + S_1 = 20 \tag{3.1}$$

$$2x_1 + 5x_2 \geq 14 \text{ would turn into } 2x_1 + 5x_2 - S_2 = 14 \tag{3.2}$$

Another important preliminary notion is that of whether the all-slack basis is a feasible departure point. The concept of all-slack basis will be explained later; for now, just imagine you are trying to find out if Equations 3.1 and 3.2 hold for values of $x_1 = 0$ and $x_2 = 0$. As you can probably verify for Equation 3.1, they do hold but not for Equation 3.2. Look now at the problem formulated in Equations 3.3 and 3.4 to illustrate the simplex method, and verify if such equations hold when you take values of zero for x_1 and x_2. As it turns out, they do hold (i.e. 0 is indeed smaller than 20, and 0 is indeed smaller than 12).

$$4x_1 + 3x_2 \leq 20 \tag{3.3}$$

$$3x_1 + x_2 \leq 12 \tag{3.4}$$

3.3.2 Equation-Based Approach

Any of the simplex methods presented in this chapter or anywhere else follow a common approach: it departs from a starting point and follows a given direction until no further improvement can be achieved. The simplex apparatus engine is conformed by a process that moves variables in and out of a set called the basis. I like to think of such a set as the equivalent of a very exclusive club, in which only wealthy VIP members can enter. As such, the basis is the club where those in are members with money in their pockets (variables with a value different than zero) and those out are the poor guys (like me) with no money (zero) in their pockets.

The initial point is commonly chosen to match the moment when the club just opens its doors to the public (say, on a Friday night); at that moment, everybody is out, that is, all decision variables have a value of zero. From that moment on, people (variables) will attempt to enter the club (the basis), but admittance is limited to one at a time. The problem turns into who is the best candidate to be admitted; as it turns out, it will be solely based upon the capability of the candidate to contribute to the party environment in the club (improvement in the objective function).

This selection is guided by the coefficients in the objective function. For a maximization problem, variables accompanied by a positive coefficient are equal candidates to enter the basis (improve the objective). For a minimization problem, variables with a negative coefficient are candidates to improve the objective. If several variables fulfil this requirement, then just simply chose one arbitrary, the algorithm will take care of bringing the other variable in eventually. This variable will go into the basis, and to make things interesting, one of the variables already in the basis will be kicked out.

This is equivalent to having a club that is full, and to get in an individual, another one must go out.

The mechanism used to identify the variable abandoning the basis is based on the ratio of the limiting amount over the coefficient of the variable entering the basis. For instance, consider the case depicted in Equations 3.5 and 3.6 in addition to the non-negativity ($x_1 \geq 0$, $x_2 \geq 0$). If the objective is given by Equation 3.7, then x_1 is definitely the only variable capable of improving the value of the objective once it takes any positive value (remember the non-negativity condition previously described). Hence, x_1 is going into the basis. Note that Equations 3.5 and 3.6 are given in the format of our first basis – the all-slack basis:

$$S_1 = 20 - 4x_1 - 3x_2 \tag{3.5}$$

$$S_2 = 12 - 3x_1 - x_2 \tag{3.6}$$

$$MAX\ 3x_1 - 2x_2 \tag{3.7}$$

Ratios for the limiting amount over the coefficient of x_1 are shown on the second column of Table 3.2; the smallest non-negative of such ratios will be our indication of the variable that abandons the basis.

The final step of the simplex apparatus is a variable replacement to express all known constraints and the objective in terms of the variable that is abandoning the basis; hence, we use an expression of the variable entering the basis in terms of the variable abandoning the basis (Equation 3.9) to make such replacements. Equations 3.8 and 3.10 show such replacement on the other constraint and the objective function:

$$S_1 = 20 - 4x_1 - 3x_2 \tag{3.8}$$

$$x_1 = 4 - 1/3x_2 - 1/3S_2 \tag{3.9}$$

$$MAX\ 3(4 - 1/3x_2 - 1/3S_2) - 2x_2 = 12 - 3x_2 - S_2 \tag{3.10}$$

At this stage, we return to the question of improvement in an attempt to improve the objective: looking at the coefficients of the variables at the objective function, we wonder which variable could help on the quest of a higher value, in this particular case no one, and then the problem has terminated.

TABLE 3.2

Basis and Ratios

Constraint	Ratio
$4x_1 + 3x_2 + S_1 = 20$	$20/4 = 5$
$3x_1 + x_2 + S_2 = 12$	$12/3 = 4$

If a variable were able to improve the objective, then we would have calculated ratios to identify the variable that abandons the basis and proceed through another cycle of variable replacements. However, that is not the case in this simple example, and at this stage, it is clear that $x_1 = 4$ and the objective takes the value of 12. Notice that neither x_2 nor S_2 is in the basis, meaning their values are zero.

In summary:

1. Expand the problem and add slack variables.
2. Verify if the all-slack basis is a feasible basis (simply check if the inequalities hold when all decision variables are equal to zero).
3. Choose a variable to enter the basis (look at the coefficients of your current objective).
4. Select a variable to abandon the basis: for this, prepare ratios (fractions) by dividing the limiting amount of each equation over the coefficient of the variable going into the basis (previous step), and choose the smallest non-negative to enter the basis.
5. Express the variable entering the basis in terms of the one exiting the basis using the corresponding constraint.
6. Replace the variable found in 3.7 at all constraints and the objective.
7. Repeat from 3.5 to 3.8.
8. Termination criteria: the objective cannot be further improved.

   Note: The sense of improvement is conditional on the values that the decision variables can take; in this case, only positive values guide our example.

3.3.3 Tableau Approach

The simplex tableau approach follows the same principles summarized earlier. However, it relays upon a table format in order to avoid the unnecessary use of the variables. It concentrates its attention to manipulate the values of the coefficients accompanying the variables of the constraints and the objective function. If one more caveat is required, the sense of improvement will be reversed. That is, for a maximization problem, we will consider that negative coefficients do help to improve and positive do not. For this to be feasible, we will change the signs on the objective (we will multiply all by -1).

Let us look at the previous example one more time. First, change the sign of the coefficients at the objective function. Then expand the constraints with slack variables. Table 3.3 presents the data used in the creation of our first tableau in Table 3.4. At this stage, it should be evident that the tableau is a more organized and faster way to proceed than the equation-based approach presented before.

TABLE 3.3

Preparing for the Tableau

Constraint	Ratio
$4x_1 + 3x_2 + S_1 = 20$	$20/4 = 5$
$3x_1 + x_2 + S_2 = 12$	$12/3 = 4$
$MAX - 3x_1 + 2x_2$	NA

TABLE 3.4

Departure Tableau

Basis	x_1	x_2	S_1	S_2	Ratios
S_1	4	3	1	0	$20/4 = 5$
S_2	3	1	0	1	$12/3 = 4$
MAX	3	2	0	0	Not applicable

TABLE 3.5

Normalization on the Pivot Column by the Pivot Element

Basis	x_1	x_2	S_1	S_2	Value
S_1	4	3	1	0	20
x_1	1	1/3	0	1/3	4
MAX	3	2	0	0	0

As before, from the objective we identify that x_1 will enter the basis and from the ratios we find that S_2 will leave us. The actual variable replacement from before will be replaced by a pivot mechanism. The mechanism starts by changing the variable name in the basis column, and in this case, S_2 will be replaced by x_1. Then a normalization of the pivot row follows. The pivot row is that of the variable abandoning the basis, and the pivot column is that of the variable entering the basis. We normalize the pivot row by dividing all coefficients by the pivot element (the one at the intersection of pivot's row and column). Table 3.5 shows this first step (bear in mind zero remains zero after the normalization).

The second step takes all basic variables columns and replaces them by identities only (i.e. it allocates values of 1 and 0). In our case, S_1 and x_1 are basic variables, so we will do that for them. The only coefficient that changes is that of the first equation for x_1, the 4 is replaced by a 0, all other amounts remain the same, and it is easy to verify that we have identities for the columns of S_1 and x_1 in Table 3.6.

The last step works out the columns of the non-basic variables following the formula. Basically, it take the current value of each cell and subtract the

TABLE 3.6

Identities for Basic Variables

Basis	x_1	x_2	S_1	S_2	Value
S_1	0	3	1	0	20
x_1	1	1/3	0	1/3	4
MAX	0	2	0	0	0

TABLE 3.7

Simplex Pivot

Basis	x_1	x_2	S_1	S_2	Value
S_1	0	3–4(1/3)	1	0–4(1/3)	20 – 4(4)
x_1	1	1/3	0	1/3	4
MAX	0	2–3(1/3)	0	0–3(1/3)	0–3(4)

TABLE 3.8

End of Simplex Pivot

Basis	x_1	x_2	S_1	S_2	Value
S_1	0	5/3	1	−4/3	4
x_1	1	1/3	0	1/3	4
MAX	0	−1	0	−1	−12

multiplication of the corresponding value of the original pivot column (before normalizing) multiplied by the normalized value of the corresponding pivot row. This operation does not apply to the pivot row but does apply to the values and the objective function. Table 3.7 illustrates the detailed calculations for our example.

Solving for this simple arithmetic operations provides us with the end of our first simplex pivot (and as we know for this case also of the entire optimization process). As seen, both objectives in Table 3.8 are negative and no further improvement can be achieved. Also note that x_2 and S_2 are both zero. From Table 3.8, it is easy to see that $S_1 = 4$, $x_1 = 4$ and the objective is 12. It is important here to mention that the sign for the value of the objective should always be changed when following the approach herein presented. Alternatively, some books change the value of the coefficients of the objective at the beginning and reverse the sense of improvement; however, my students always find that difficult to understand and I prefer to simply do a sign change at the end instead of having to argue for such alternate approach.

In summary,

1. Expand the problem and add slack variables. Transfer the coefficients to the tableau (the table).
2. Verify if the all-slack basis is a feasible basis (simply check if the inequalities hold when all decision variables are equal to zero).
3. Choose a variable to enter the basis: if maximizing, take any of the variables with a positive coefficient at the objective (Z) row (this gives you your pivot column).
4. Select a variable to abandon the basis: smallest non-negative ratio (value over coefficient) of each basic variable (this gives you your pivot row).
5. Create a new table and write the normalized pivot row (divide old pivot row by pivot element value).
6. Make identities for all basic variables (one on the intersection of a variable and itself, zero everywhere else on the column of a basic variable).
7. Do a pivot operation for the remaining non-basic variable column values: old cell value ***minus*** (corresponding pivot column ***times*** corresponding pivot row of this new tableau).
8. Repeat steps 3–7.
9. Termination criteria: the objective cannot be further improved.

   Note: The sense of improvement is conditional on the values that the decision variables can take; in this case, only positive values guide our example. The pivot element is found at the intersection of the pivot row and pivot column.

3.4 Simplex Method: Other Considerations

3.4.1 Alternate Optima

It is important to recall that the type of problems solved by the simplex algorithm is linear in nature and that sometimes the objective exhibits the same slopes as one of the constraints (linearly dependent); in this event, the solution will be characterized by alternate optima, that is, there will be more than one combination of decision variables yielding the optimal value for the objective. The way to identify alternate optima is easy: for our equation-based approach or tableau-format procedure, we will find an objective after a pivot operation with a value of zero for one of the (non-basic) variables. In our example, the coefficient of either x_2 or S_2 will be zero instead of the -1 found in Table 3.8.

How do we deal with such situation? Simple: repeat one more pivot by choosing the non-basic variable with a coefficient of zero to enter the basis; the new values obtained will be the alternate solution to your problem. Hence, the problem will have two possible solutions.

3.4.2 Unbounded and Infeasible

Once we tackle problems in multidimensional space, how do we identify if the problem is unbounded or infeasible? Remember an unbounded problem only requires us to look again at the formulation and provide a limiting amount for the unbounded variable (resource), but an infeasible problem poses a more challenging issue because the very own formulation needs to be revised to fix deeper problems that lay on it.

An unbounded problem is characterized by all values of the ratios being either negative or infinity, meaning nobody can go into the basis. This happens when nothing can limit the amounts of interest in the direction of movement. Remember: when you select the variable coming into the basis, what you are actually doing is choosing the direction in which your algorithm will walk on the feasible space. When you chose the variable abandoning the basis, you are selecting how many steps (limiting amount) you can walk in such direction.

An infeasible problem will be identified if we land at a situation in which we have a negative value for any of the basic variable coefficients, meaning that the slack variable does not work (remember a slack cannot take on negative values), and no further improvement is possible, resulting in the inability to scape this undesirable situation, so we are basically stuck into this situation and cannot get out of it.

To fix an infeasible problem, you need to revisit your formulation and fix an incorrect definition of the problem itself (if there is some). Sometimes, the very definition of constraints turns a problem infeasible and there is nothing you can do about it. That is, given the amount of resources, you cannot accomplish what the objective is asking for. This happens in more specialized problems in which you are using a very strong constraint such as the need to improve every period a given variable by some percentage, for example, think of the desire of the department of transportation to reduce every year 10% of the total amount of roads in poor condition with a given budget. Sometimes, it simply happens that given initial circumstances of the composition of the network and available interventions and their corresponding cost, it becomes infeasible either to attain the goal during the initial periods or to sustain all roads that had turned into good condition; hence, it is the nature of the problem, the one that results in the infeasibility. We will analyse a simplified version of this problem in Chapter 8 in which we show some real-world applications to civil engineers in their day-to-day work.

3.5 Simplex Method: Examples

We now turn our attention to several guided examples that will illustrate more realistic applications of the simplex equation-based and tableau-format approaches. If you are solely interested in mastering the simplex method, I suggest you skip the description and jump directly into the formulated problem and the steps to solve it. However, I strongly encourage you to follow the examples as they will aid your problem formulation and modelling skills. We start with some solvable exercises and then include problems where alternate optima can be found. We finalize this section with examples that illustrate unbounded and infeasible problems.

3.5.1 Example 1: A Municipal Engineer in Trouble (Infeasible)

Let us start with a simple example of a municipality that is replacing water meters (the machine that measures how much water each house consumes per month). As a municipal engineer, you are tasked with replacing the entire stock of water meters (10,000 houses), and for this, you need to use the public works department personnel (the foreman is not very happy and is not willing to give you many employees). You can alternatively/complementarily hire external workers to do the task. Hence, your objective is to minimize the total cost and to achieve the task at hand (replacement of water meters).

Call x_1 the number of public works employees (internal workers) you will use and x_2 the number of external workers you will hire. To avoid major disturbances to the public, the town council had asked you to limit replacements to 20 per day. A skilled external worker can replace 5 m per day and a skilled internal worker can replace 4 m per day. You pay \$10 per internal worker per day and \$12 per external worker per day.

Finally, the legislation forces you to use at least twice the number of internal employees per external employee you hire per month (which equates to a daily basis in the same number). This constraint looks like $x_1 \geq 2x_2$ which once you move terms to one side result into $x_1 - 2x_2 \geq 0$ or alternatively $0 \geq 2x_2 - x_1$. How many workers should you hire? Table 3.9 sets the problem in the linear programming language, as you will later see such absurd regulation will make the problem infeasible.

It is important to notice that if we take $x_1 = 0$ and $x_2 = 0$, both constraints will be satisfied (simply replace their values and verify that the inequality is satisfied), and therefore the origin is a feasible departure point, that is, the problem will start from a basis with only slack variables (the all-slack basis). Either decision variable can take on a value of zero or positive. The ideal value of this objective is the trivial departure point, and therefore, the problem is ill conditioned, if we keep it under the current formulation, because we are looking for negative coefficients and there is none, so the simplex algorithm

TABLE 3.9

Problem Formulation for Example 1

Element	Composition	Units
Replacement	$4x_1 + 5x_2 \leq 20$	Water meters per day
Legislation	$-x_1 + 2x_2 \leq 0$	Workers ratio per month = day
Minimize	$10x_1 + 12x_2$	Dollars per day

Basis	x_1	x_2	S_1	S_2	Limiting Value
S_1	4	5	1	0	20
S_2	−1	2	0	1	0
MIN	10	12	0	0	0

TABLE 3.10

Problem Formulation Modified for Example 1

Element	Composition	Units
Labour cost	$10x_1 + 12x_2 \leq 120$	Dollars per day
Legislation	$-x_1 + 2x_2 \leq 0$	Workers ratio per month = day
Maximize	$4x_1 + 5x_2$	Replacements per day

TABLE 3.11

Departure Tableau Modified

Basis	x_1	x_2	S_1	S_2	Limiting Value
S_1	10	12	1	0	120
S_2	−1	2	0	1	0
MAX	4	5	0	0	0

is already at an optimal point $x_1 = 0$ and $x_2 = 0$. We change the formulation (Table 3.10) in an attempt to use the simplex method to solve: the constraint for S_1 becomes our new objective; for this, imagine that we want to achieve as many replacements as possible while not exceeding a given budget of, say, \$120 in labour cost per day and the replacement rate is the new objective to be maximized (Table 3.11).

Again, the all-slack basis is feasible, hence the departure point is $x_1 = 0$ and $x_2 = 0$. Our new objective is to maximize and hence either decision variable will aid to improve the objective. We chose to start with x_1 (you can start with x_2 as practice and should arrive to the same answer). Looking at the ratios, we realize the smallest non-negative is actually zero, hence S_2 leaves the basis (Table 3.12). The pivot row is that of S_1 and the pivot column is that of x_1.

TABLE 3.12

Ratios (x_1 in, S_1 out)

Basis	x_1	x_2	S_1	S_2	Ratios
S_1	10	12	1	0	120/10 = 12
S_2	−1	2	0	1	0/−1 = 0
MAX	4	5	0	0	0

TABLE 3.13

Normalization on the Pivot Row by the Pivot Element

Basis	x_1	x_2	S_1	S_2	Ratios
x_1	1	12	1	0	120
S_2	0	1	0	1	0
MAX	0	5	0	0	0

TABLE 3.14

Identities for the Basic Variables

Basis	x_1	x_2	S_1	S_2	Ratios
x_1	1	12	1	0	120
S_2	0	1	0	1	0
MAX	0	5	0	0	0

TABLE 3.15

Pivot Operation (x_1 in, S_1 out)

Basis	x_1	x_2	S_1	S_2	Ratios
x_1	1	12	1	0	120
S_2	0	1 − (−1)(1)	0 − (−1)(1)	1	0 − (−1)(120)
MAX	0	5 − 4(1)	0 − (4)(1)	0	0 − 4(120)

Normalization in this case requires a division by one and so nothing changes (Table 3.13). Finally, we apply a pivot on the non-basic variables by taking each cell value minus the corresponding pivot column times the normalized pivot row value.

If the previous result is accurate, then we are facing an infeasible problem. Notice S_2 has taken on a negative value (Tables 3.14 through 3.20).

TABLE 3.16

Simplex Pivot

Basis	x_1	x_2	S_1	S_2	Ratios
x_1	1	12	1	0	120
S_2	0	2	1	1	120
MAX	0	1	−4	0	−480

TABLE 3.17

Ratios (x_2 in, x_1 out)

Basis	x_1	x_2	S_1	S_2	Ratios
x_1	1	12	1	0	$120/12 = 10$
S_2	0	2	1	1	$120/2 = 60$
MAX	0	1	−4	0	−480

TABLE 3.18

Identities and Normalization by 1

Basis	x_1	x_2	S_1	S_2	Ratios
x_2	1	1	1	0	120
S_2	0	0	1	1	120
MAX	0	0	−4	0	−480

TABLE 3.19

Pivot Operation for Non-Basic (x_1, S_1)

Basis	x_1	x_2	S_1	S_2	Ratios
x_2	1	1	1	0	120
S_2	$0 - 2(1)$	0	$1 - 2(1)$	1	$120 - 2(120)$
MAX	$0 - 1(1)$	0	$-4 - 1(1)$	0	$-480 - 1(120)$

TABLE 3.20

Pivot Operation for Non-Basic (x_1, S_1)

Basis	x_1	x_2	S_1	S_2	Ratios
x_2	1	1	1	0	120
S_2	−2	0	−1	1	−120
MAX	−1	0	−5	0	−600

3.5.2 Example 2: A Construction Materials Company

Let us continue with a relatively simple example of a materials company that produces flooring of three types: residential (x_1), commercial (x_2) and industrial (x_3). Each floor category has assigned a given price (p_1, p_2, p_3). The company has to pay for its labour, a wage per hour w, which is the same for all three types of floors and must rent the use of three specialized machines at rates r_1, r_2, r_3 per hour; this rate is net. The problem of this company as of any other company in the classical literature of economics is to maximize profits.

Profits are given as the revenue for sales minus the total cost. The revenue comes from the amount of goods produced times their price; cost comes from two main sources: (a) labour cost from wages times the number of hours of work and (b) capital cost from the number of hours of machine used for production times the rate of return minus the depreciation rate, which in this case has been represented directly as an effective rental rate of machines. Table 3.21 summarizes the problem information and gives some specific numbers to it. Before that, it is important to clarify that a competitive market means two things for us: first, that prices are given (that is fixed to us and we cannot change them), and second, that everybody charging these prices for their flooring can sell as many units as the market requires. This comes from the fact that you are charging what people are willing to pay, and hence you can sell all the units that you produce assuming there is a large enough market to not impose a limit on your sales. I have given values to prices and wage and rental rates and now the question turns into how many units to produce.

If you recall from previous chapters, we were provided with values of coefficients in the objective with no explanation of their origin; as you can see, the values of any objective do come from a calculation, and in this particular case, they will be given by the price minus the wage minus the rental rate. Table 3.22 provides a sample of values we will use for the remaining of this example. Remember, the generic formulation previously provided is only one of many ways we determine the weights of an objective, and each problem type will have very specific elements conditioning such calculation.

Now is the time to look at the constraints. For this problem, assume you have a fixed amount of storage on your warehouse which limits your maximum daily production (each morning trucks come before the production

TABLE 3.21

Determining the Objective: Generic Formula

Element	Composition
Revenue	$p_1x_1 + p_2x_2 + p_3x_3$
Labour cost	$wx_1 + wx_2 + wx_3$
Capital cost	$r_1x_1 + r_2x_2 + r_3x_3$
Profit	$(p_1 - w - r_1)x_1 + (p_2 - w - r_2)x_2 + (p_3 - w - r_3)x_3$

TABLE 3.22

Determining the Objective: Sample Values

Element	Composition
Revenue	$5x_1 + 7x_2 + 9x_3$
Labour cost	$2x_1 + 2x_2 + 2x_3$
Capital cost	$1.2x_1 + 1.5x_2 + 1.8r_3x_3$
Profit calculation	$(5 - 2 - 1.2)x_1 + (7 - 2 - 1.5)x_2 + (9 - 2 - 1.8)x_3$
Final profit	$1.8x_1 + 3.5x_2 + 5.2x_3$

starts and all goods produced are shipped to their final destination; we will actually analyse this problem in a later chapter). In this case, we will impose an overall constraint to the combined amount of units from x_1, x_2 and x_3, but we could have had one per each of them. A maximum of 50 units from either of the types of flooring can be stored per day (Table 3.23).

Another constraint comes from the productivity of labour, a similar calculation to that of the objective applies and values of productivity are determined by industrial engineers. For all you know, it takes 0.5 hour to produce one unit of x_1, 1.5 hours to produce one unit of x_2 and 2 hours to produce one unit of x_3. Total labour in the plant is given by 10 employees allocated to the production of any of these goods but that do follow union laws and cannot work more than 40 hours per week, or 8 hours per day.

Table 3.24 shows the final formulation for this problem, and we have two constraints and one objective; hence, two slack variables are required (one per constraint). Notice that your decision variables are the number of units of each of the three types of flooring which cannot take negative values. This signifies that the non-negativity constraint applies and that the all-slack basis

TABLE 3.23

Determining the Constraints

Element	Composition	Units
Storage	$x_1 + x_2 + x_3 \leq 50$	Units of floor per day
Labour	$0.5x_1 + 1.5x_2 + 2x_3 \leq (10)8 = 80$	Hours per day

TABLE 3.24

Problem Formulation

Element	Composition	Units
Storage	$x_1 + x_2 + x_3 \leq 50$	Units of floor per day
Labour	$0.5x_1 + 1.5x_2 + 2x_3 \leq 80$	Hours per day
Maximize	$1.8x_1 + 3.5x_2 + 5.2x_3$	Dollars per day

TABLE 3.25

Departure Tableau

Basis	x_1	x_2	x_3	S_1	S_2	Limiting Value
S_1	1	1	1	1	0	50
S_2	0.5	1.5	2	0	1	80
MAX	1.8	3.5	5.2	0	0	0

is in the feasible space (what!). This simply means that the departure point for our walking simplex algorithm will start at $x_1 = 0, x_2 = 0, x_3 = 0$ (the origin if you want). The departure tableau is shown in Table 3.25 which has been placed next to the formulation so that the equivalence can be easily seen. Our constraint for storage will be given the slack variable S_1 and our constraint for labour S_2, the departure basis is given by an all-slack basis (i.e. S_1 and S_2) and coefficients from each basic and non-basic variable are taken from the original constraints.

3.6 Simplex Method: Auxiliary Variable

So far, we have concentrated our attention to problems for which the all-slack basis is in the feasible space. However, nothing has been said as to how to proceed when this is not the case.

If the all-slack basis is not located in the feasible space, we need to find a departure point that is indeed in the feasible space. Given that the simplex method is a walking algorithm that visits the feasible space, we can still use it to find the first feasible point. However, to achieve that we want the method to find the closest point to the origin (all-slack basis) that is in the feasible space. For this reason, we will minimize the distance to be walked; otherwise, we could risk the method moving in the right direction but passing over the feasible space.

A variable called 'auxiliary' will help us achieve this. The value of this variable is to be minimized in order to satisfy the earlier argument. Our objective changes to such minimization, and we keep on hold the original objective of the formulation for a later stage.

The auxiliary variable is included on each constraint, and it enters with the opposite sign of the slack variable. The auxiliary variable is also chosen (always) arbitrarily to be the first variable to enter the basis. The first variable to exit the basis will be the one from the largest non-holding constraint (that for which the all-slack basis is not compliant).

Consider, for example, an extended version of the example presented in Equations 3.1 and 3.2, with the addition of two more constraints shown in

Equations 3.11 through 3.14. Note that the same auxiliary variable enters each and every equation as opposed to independent slack variables per equation:

$$3x_1 + 2x_2 \leq 20 \text{ would turn into } 3x_1 + 2x_2 + S_1 - A = 20 \quad (3.11)$$

$$2x_1 + 5x_2 \geq 14 \text{ would turn into } 2x_1 + 5x_2 - S_2 + A = 14 \quad (3.12)$$

$$4x_1 + 4x_2 \geq 16 \text{ would turn into } 4x_1 + 4x_2 - S_3 + A = 16 \quad (3.13)$$

$$5x_1 + 2x_2 \leq 100 \text{ would turn into } 5x_1 + 2x_2 + S_4 - A = 100 \quad (3.14)$$

The system of equations described before needs to be solved for an objective that is simply the minimization of A (*MIN A*). Either the equation-based approach or the tableau-format method could be chosen to accomplish such a task. Table 3.26 illustrates the tableau-format method set-up. Note that a constraint that does not contain a particular slack is equivalent to a zero multiplying such a slack variable.

For this example, the largest conflicting constraint is the one for S_3, so in our first step, the auxiliary variable will enter the basis and S_3 will abandon the basis. Our pivot row is that of S_3 where the label for the basis gets replaced by A and its elements normalized by 1 (so they do not change) (Table 3.27 shows this step). The pivot column is that of A and will be used soon. The second step (after normalization) is that of identities for the basic variables, in this case for S_1, S_2, A and S_4 as shown in Table 3.28. Finally, we have to conduct a pivot operation for the non-basic variables (i.e. x_1, x_2, S_3 as well as

TABLE 3.26

End of Simplex Pivot

Basis	x_1	x_2	S_1	S_2	S_3	S_4	A	Value
S_1	3	2	1	0	0	0	−1	20
S_2	2	5	0	−1	0	0	1	14
S_3	4	4	0	0	−1	0	1	16
S_4	5	2	0	0	0	1	−1	100
MAX	0	0	0	0	0	0	1	0

TABLE 3.27

Normalization of Pivot Row

Basis	x_1	x_2	S_1	S_2	S_3	S_4	A	Value
S_1	3	2	1	0	0	0	−1	20
S_2	2	5	0	−1	0	0	1	14
A	4	4	0	0	−1	0	1	16
S_4	5	2	0	0	0	1	−1	100
MAX	0	0	0	0	0	0	1	0

TABLE 3.28
Identities

Basis	x_1	x_2	S_1	S_2	S_3	S_4	A	Value
S_1	3	2	1	0	0	0	0	20
S_2	2	5	0	1	0	0	0	14
A	4	4	0	0	−1	0	1	16
S_4	5	2	0	0	0	1	0	100
MAX	0	0	0	0	0	0	0	0

TABLE 3.29
Pivot Operation

Basis	x_1	x_2	S_1	S_2	S_3	S_4	A	Value
S_1	3 − 1(4)	2 − 1(4)	1	0	0 − 1(−1)	0	0	20 − 1(16)
S_2	2 − 1(4)	5 − 1(4)	0	1	0 − 1(−1)	0	0	14 − 1(16)
A	4	4	0	0	−1	0	1	16
S_4	5 − 1(4)	2 − 1(4)	0	0	0 − 1(−1)	1	0	100 − 1(16)
MAX	0 − 1(4)	0 − 1(4)	0	0	0 − 1(−1)	0	0	0 − 1(16)

TABLE 3.30
End of Simplex Pivot

Basis	x_1	x_2	S_1	S_2	S_3	S_4	A	Value
S_1	7	6	1	0	−1	0	0	36
S_2	−2	1	0	1	1	0	0	14
A	4	4	0	0	−1	0	1	16
S_4	9	6	0	0	−1	1	0	116
MAX	−4	−4	0	0	1	0	0	−16

for the value and the objective function) in Table 3.29 to obtain the end of this pivot operation (Table 3.30).

Recall that we are minimizing, and hence any coefficient with negative symbol will help achieve this; therefore, either x_1 or x_2 would serve for this purpose. Arbitrarily, we chose x_1 to enter the basis.

3.7 Multiple Objectives and Trade-Off Analysis

To this point, you have probably noticed that all previous analysis and examples had made use of one objective. However, many problems in real life involve several objectives and require some sort of comparison among alternatives from possible combinations of the objectives. This chapter explains two simple methods to deal with several objectives and various approaches to compare alternatives, that is, to conduct trade-off analysis.

Real-life problems involve multiple, often conflicting, objectives. Consider, for instance, your very own case: you want to obtain good grades in your courses while *enjoying life* (eating, socializing and hanging out with your friends). Maximizing your grades requires you to invest time, call n_t the amount of time you dedicate to study; it turns that $1 - n_t$ is the proportion of your time available to *enjoy life*; allow me to use $u(1 - n_t)$ to represent the total utility from such banal activities.

Hence, a maximization of grades requires the usage of n_t, with an upper bound given by 1 (or 100 %); however, maximizing utility (enjoying life) also requires the usage of n_t, thereby resulting in a conflict for the allocation of n_t. One way to handle two objectives is by turning one of them into a constraint.

A couple of examples are used to illustrate conflicting objectives and how to turn one of them into a constraint. Section 3.7.1 explores the solution for the work/study problem. Section 3.7.2 illustrates the conflicting objectives behind road maintenance and rehabilitation.

3.7.1 Example of Conflicting Objectives: Maximizing Lifetime Income

An individual faces the problem of maximizing his lifetime income (Equation 3.15): income comes from wages (w_t), and they depend on the amount of skills or human capital (h_t) that the individual had acquired through education and the amount of time that the individual dedicates to work ($1-n_t$); however, the individual can also use time to acquire more human capital by dedicating a fraction of her time to acquire more skills and this results into higher prestige and income, so he or she would want to maximize it as well (Equation 3.16).

This problem is a simplified version of the one presented by Porath 1976. In order to avoid dealing with two objectives, we can turn the second objective into a constraint, and in this manner, we eliminated the problem of dealing with two objectives: as shown in Equations 3.18 through 3.20, the second objective has been equalized to a future amount called h_{t+1} which simply represents the new level of skills on the next time period:

$$\max \sum w_t(1 - n_t)h_t \tag{3.15}$$

$$\max(1 - d)h_t + z_t(n_t h_t)^{g1} \tag{3.16}$$

$$h_{t+1} = (1 - d)h_t + z_t(n_t h_t)^{g1} \tag{3.17}$$

3.7.2 Maintenance and Rehabilitation: An Example of Conflicting Objectives

The minister of transportation needs to schedule the maintenance and rehabilitation of roads within the province which is guided by a binary decision variable $x_{t,i,j}$ that takes on the value of one whenever on year t segment i receives intervention j (otherwise, zero). The minister wishes to maximize the level of condition (Q) which depends on previous year's condition Q_{t-1} and can either deteriorate D or improve I (Equation 3.18).

The minister also wishes to minimize the overall cost which is the result of multiplying unitary cost c_{tij} on year t of treatment j on asset i times the binary decision variable as shown in Equation 3.19. To improve the condition, he needs to schedule maintenance and rehabilitation interventions, and to pay for them; hence, both objectives are conflicting. Just like in the previous example, we turn the second objective into a budget constraint by simply limiting total cost to a given budget B_t per period of time t:

$$\max \sum L_i[x_{tij}(Q_{t-1} + I) + (1 - x_{tij})(Q_{t-1} - D)] \tag{3.18}$$

$$\min \sum c_{t,i,j} L_i x_{t,i,j} \tag{3.19}$$

$$\sum c_{t,i,j} L_i x_{t,i,j} <= B_t \tag{3.20}$$

In some special cases, one can replace the expression of one objective into the other one, but this is limited to few cases. The most common manner to handle several objectives at the same time is that of the weighted grand objective which consist of a simple aggregation of terms.

However, it is important to respect three considerations: (1) use a minus sign when adding a minimizing objective to a maximizing objective, (2) bare in mind objectives could not be on the same units and some sort of scaling may be required and (3) use relative weights to represent relative importance of each objective.

Let us consider the case of Section 3.7.1: in this case, the first objective is given in dollars and the second one in units of human capital (or skills); adding both objectives is tricky, and a rescaling to a common scale may be required.

A similar case happens for Section 3.7.2: the first objective is given in units of condition and the second one in units of money (dollars). A common way to scape this dilemma is by arbitrarily alternating complementary weights to each objective and conducting a trade-off analysis of inferiority. Another approach is simply to solve both objectives and attempt to map the feasible space.

3.8 Global Objective: Alternating Weights

One of the most common ways to face two conflicting objectives is by joining them into one grand objective. A special case is that in which both objectives share the same decision variables. Before adding two objectives, it is important to identify if both are given the same importance or not. Some authors consider this as a bias, while others allow for user's criteria (preferences) to come to the equation in the form of weights.

If you agree with the need to do not involve your subjective criteria, then the best you can do is to come up with a set of alternating weights that allow you to map all possible combinations of both weights. If not, then you simply use your own preferences to assign upfront a weight before proceeding with the summation.

In what follows, we use alternating weights to solve for n possible combinations of the objectives. Notice that this means that the slope of the objective line will rotate and a set of optimal points will be identified. An additional implication is that those corner points of the feasible space not visited by the objective line are ruled off by the analysis. Let us look at a couple of examples.

3.8.1 Alternating Weights: An Example of Floor Production

Consider the production of ceramic tiles from a previous chapter, the original objective is profit maximization given by $140x_1 + 160x_2$; let us add a new objective given by the amount of gas emissions (CO_2).

Presume each unit of x_1 produces 500 g per year of CO_2 equivalent gas emissions and each unit of x_2 produces 200 g per year. The new objective will be to minimize $500x_1 + 200x_2$. We are going to alternative weights using 0.1 increments/decrements. Table 3.31 shows the combinations.

We used the variables a_1 and a_2 to denote the relative weights for the first and second objectives. The column global objective illustrates the summation of both objectives after multiplying them by the corresponding weights. The only remaining step is to solve for the new objective, and we leave this as an exercise for the reader. You simply need to use the constraints and solve for each specific global objective.

TABLE 3.31

Assembling of the Global Objective

a_1	a_2	Global Objective
0.9	0.1	$0.9(140x_1 + 160x_2) + 0.1(500x_1 + 20x_2)$
0.8	0.2	$0.8(140x_1 + 160x_2) + 0.2(500x_1 + 20x_2)$
0.7	0.3	$0.7(140x_1 + 160x_2) + 0.3(500x_1 + 20x_2)$
0.6	0.4	$0.6(140x_1 + 160x_2) + 0.4(500x_1 + 20x_2)$
0.5	0.5	$0.5(140x_1 + 160x_2) + 0.5(500x_1 + 20x_2)$
0.4	0.6	$0.4(140x_1 + 160x_2) + 0.6(500x_1 + 20x_2)$
0.3	0.7	$0.3(140x_1 + 160x_2) + 0.7(500x_1 + 20x_2)$
0.2	0.8	$0.2(140x_1 + 160x_2) + 0.8(500x_1 + 20x_2)$
0.1	0.9	$0.1(140x_1 + 160x_2) + 0.9(500x_1 + 20x_2)$

3.8.2 Solving Alternating Weights: A Brief Example

Let us consider the example shown in Equations 3.21. Table 3.32 shows the formulation which follows the same steps of the previous example. Table 3.33 is an expanded version showing in the last two columns the values of the global objective after solving the system of equations with alternating weights (a_1 and a_2).

$$\begin{aligned} &\max 3x_1 - 2x_2 \\ &\max -x_1 + 2x_2 \\ &3x_1 - 6x_2 \leq 6 \\ &4x_1 - 2x_2 \leq 14 \\ &x_1 \leq 6 \\ &-x_1 + 3x_2 \leq 15 \\ &-2x_1 + 4x_2 \leq 18 \\ &-6x_1 + 3x_2 \leq 9 \end{aligned} \tag{3.21}$$

As seen in the analysis, it identifies four possible points which become candidates for the optimal solution (10, −2), (8, 4), (4, 8), (−3, 9).

Figure 3.4 illustrates the feasible space; as seen, the upper-right region encompasses the optimal set from which one point will be chosen. Very soon we will cover various methods that aid the decision maker on how to select a solution among these non-inferior alternatives.

TABLE 3.32

Formulation of the Global Objective

a_1	a_2	Global Equation
1	0	$3x_1 - 2x_2$
0.9	0.1	$2.6x_1 - 1.6x_2$
0.8	0.2	$2.2x_1 - 1.2x_2$
0.7	0.3	$1.8x_1 - 0.8x_2$
0.6	0.4	$1.4x_1 - 0.4x_2$
0.5	0.5	x_1
0.4	0.6	$0.6x_1 + 0.4x_2$
0.3	0.7	$0.2x_1 + 0.8x_2$
0.2	0.8	$0.2x_1 + 1.2x_2$
0.1	0.9	$0.6x_1 + 1.6x_2$
0	1	$-x_1 + 2x_2$

TABLE 3.33
Solving the Global Objective

a_1	a_2	Global Equation	$maxz_1$	$minz_2$
1	0	$3x_1 - 2x_2$	10	−2
0.9	0.1	$2.6x_1 - 1.6x_2$	10	−2
0.8	0.2	$2.2x_1 - 1.2x_2$	10	−2
0.7	0.3	$1.8x_1 - 0.8x_2$	8	4
0.6	0.4	$1.4x_1 - 0.4x_2$	8	4
0.5	0.5	x_1	8	4
0.4	0.6	$0.6x_1 + 0.4x_2$	4	8
0.3	0.7	$0.2x_1 + 0.8x_2$	4	8
0.2	0.8	$0.2x_1 + 1.2x_2$	4	8
0.1	0.9	$0.6x_1 + 1.6x_2$	−3	9
0	1	$-x_1 + 2x_2$	−3	9

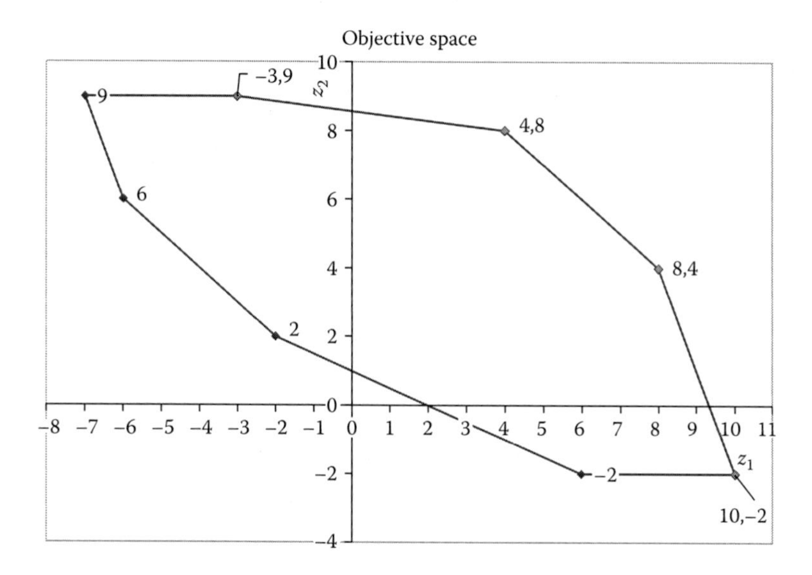

FIGURE 3.4
Graphical solution of the global objective.

3.9 Analysis of Inferiority

The analysis of inferiority in this section follows the famous *Pareto* optimality criteria which states that a non-inferior solution is the one in which it is not possible to improve in one criteria without reducing in another one.

There are two ways to be non-inferior: (1) the alternative is the absolute best in one of the criteria and (2) there is no other alternative performing equal or better in all criteria. Let us consider a set of simple examples to illustrate what we mean by a non-inferior and a dominated alternative (Table 3.33).

3.9.1 Example of Inferiority for a Bridge Replacement

Consider the problem of replacing a bridge that connects downtown with the suburbs. Table 3.34 shows available alternatives: alternative 1 demolishes the old bridge and builds the same thing again, alternative 2 builds a new bridge and leaves the current bridge for the exclusive use of non-motorized users (pedestrians, bikers, etc.), alternative 3 rehabilitates the existing bridge and alternative 4 demolishes the old one and builds a new one with a high-occupancy vehicle lane and a bus-exclusive lane.

Capacity in terms of vehicles per day is an important criterion, and currently the bridge carries 170,000 vehicles per day. The cost is another important criterion: it can be broken down into investment cost and possible toll per car. Another important criterion is construction time.

There are two ways an alternative can be classified as non-inferior, one is by being the absolute best in one criteria alone, and the other is by being relatively better than others in some of the criteria but not all.

An alternative that is superior than another one in all available criteria is deemed non-inferior, and the other is called dominated or inferior. For our example of the bridge, the one with the largest capacity will be declared non-inferior, the same for the cheapest one and the fastest construction time. Table 3.35 contains this portion of the dominance analysis.

TABLE 3.34

Example of Dominance Analysis

Alternative	Capacity	Cost (Billion)	Duration (Months)
A1	170,000	2	22
A2	170,000	2.5	36
A3	170,000	3	12
A4	185,000	2.2	24

TABLE 3.35

Example of Dominance Analysis

Alternative	Capacity	Cost (Billion)	Duration (Months)	Dominance
A1	170,000	2	22	Non-inferior
A2	170,000	2.5	36	
A3	170,000	3	12	Non-inferior
A4	185,000	2.2	24	Non-inferior

TABLE 3.36

Example of Dominance Analysis

Alternative	Capacity	Cost (Billion)	Duration (Months)	Dominance
A1	170,000	2	22	Non-inferior
A2	170,000	2.5	36	Dominated by A1
A3	170,000	3	12	Non-inferior
A4	185,000	2.2	24	Non-inferior

It only rests to analyse alternative A2, and the easiest way is to find an alternative that is capable of performing equal/better, that is, it suffices to be better in one criterion and equal in the rest. This is the case when you compare alternatives A2 with A1; A1 performs better in cost and duration and the same in capacity. Therefore, A2 is dominated by A1 and it is ruled off the analysis (Table 3.36).

3.9.2 Example of Inferiority for a Municipal Engineer

The analysis involving more alternatives follows the same steps: this example illustrates the case of the municipal engineer facing the selection of a contractor through a bid process. Typically, the selection of a contractor by a municipality to build a major piece of infrastructure does not only follow the need to spend as little as possible but to make sure that the contractor has the qualifications and experience that will guarantee a job well done.

Table 3.37 summarizes the tender criteria and received bids. As you can see aside from the cost we have requested, the contractor has to document successful construction of projects of a similar magnitude and nature, and also the contractor has been inquired about the availability of specialized equipment and qualified personnel; the final criterion was that of total construction time.

3.9.3 Example of Inferiority for Gas Emissions

Table 3.38 summarizes the profit and gas emissions for values on the vertex of the feasible space of the example in Section 3.2 for the ceramic tiles.

TABLE 3.37

Example of Dominance Analysis

Contractor	Quoted Cost	Experience	Equipment	Personnel
Contractor 1	155,000	2	22	100
Contractor 2	140,000	2	36	85
Contractor 3	158,000	3	12	90
Contractor 4	185,000	56	24	110

TABLE 3.38

Summary of Profit and Gas Emissions

Alternative	x_1	x_2	$maxz_1$	$minz_2$
A	0	0	0	0
B	8	0	1120	4000
C	8	2	1440	4400
D	6	4	1480	3800
E	2	6	1240	2200
F	0	6	960	1200

Let us conduct a dominance analysis for it. As you can see, the most environmental friendly option is to do nothing and have zero gas emissions as suggested by alternative A (however, such radical solution also returns zero profits).

From a purely profit perspective, alternative D is the best. Both alternatives are deemed non-inferior (Table 3.39). Now, let us turn our attention to alternative C; clearly D is a better option, as it performs better on both criteria giving a higher level of profit and a lower level of gas emissions. For this reason, alternative C is being dominated by alternative D (Table 3.40).

TABLE 3.39

Dominance Analysis

Alternative	x_1	x_2	$maxz_1$	$minz_2$	Dominance Analysis
A	0	0	0	0	Non-inferior
B	8	0	1120	4000	
C	8	2	1440	4400	
D	6	4	1480	3800	Non-inferior
E	2	6	1240	2200	
F	0	6	960	1200	

TABLE 3.40

Dominance Analysis

Alternative	x_1	x_2	$maxz_1$	$minz_2$	Dominance Analysis
A	0	0	0	0	Non-inferior
B	8	0	1120	4000	
C	8	2	1440	4400	Dominated by D
D	6	4	1480	3800	Non-inferior
E	2	6	1240	2200	
F	0	6	960	1200	

TABLE 3.41

Dominance Analysis

Alternative	x_1	x_2	$maxz_1$	$minz_2$	Dominance Analysis
A	0	0	0	0	Non-inferior
B	8	0	1120	4000	Dominated by E
C	8	2	1440	4400	Dominated by D
D	6	4	1480	3800	Non-inferior
E	2	6	1240	2200	Non-inferior
F	0	6	960	1200	Non-inferior

Now look at alternatives B and E; E has higher profit and lower pollution, hence B is dominated by E (Table 3.41). Take now alternative E and try to find somebody that performs better in both criteria; as you will find, this task is impossible: only F and A have lower levels of pollution but neither had higher levels of profit, hence alternative E is non-inferior.

Similarly, alternative F is non-inferior as only alternative A has lower levels of pollution but not higher profit.

3.10 Ranges, Satisfaction, and Lexicographic

Sometimes, the criteria considered to select the best option contain very minor differences, and hence, it is useful to use ranges or averages to equalize values of choices that result indifferent to the decision maker. Such ranges help to expedite the analysis (i.e. dominance or alike).

Satisfaction refers to the establishment of a minimum threshold level on a particular criterion, to sort alternatives based on such criterion and to eliminate those alternatives that fail to accomplish such minimum level as required by the decision maker.

A satisfaction criterion is user-specific and varies between decision makers. Hence, it is subjective and individuals could have rigid and flexible satisfaction requirements. In this book, we will consider strict requirements (deal breakers!).

Lexicographic analysis is a simple sorting mechanism that arranges alternatives using levels of sorting criteria: it sorts based on the most relevant criteria first and the resort choices based on a secondary (and sometimes tertiary or more) criterion. Let us illustrate the use of these criteria.

3.10.1 Example: Commuting to Work

Consider the case of a traveller trying to select the best way to commute to work each day. The traveller worries about cost, travel time and level of

TABLE 3.42

A Choice of Commute to Work

Option	Time (Hours)	Cost ($/Month)	Comfort	Dominance
1. Walk	4.5	0	3	Dominated by 2
2. Bike	1.18	0	4	Non-inferior
3. Bus	1	101	1	Dominated by 5
4. Car	0.42	300	10	Non-inferior
5. Walk–bus	0.82	101	0	Non-inferior
6. Car–metro(1)	0.84	145	8	Non-inferior
7. Car–metro(2)	1	305	9	Dominated by 4,8
8. Car–train(1)	0.92	175	7	Dominated by 6
9. Car–train(2)	1	175	7	Dominated by 6
10. Walk–train	1.25	125	6	Dominated by 2
11. Car–bus	0.67	151	5	Non-inferior

comfort of the trip because he has to do this trip twice a day 10 times a week (presuming he/she does not work on the weekends).

The criteria used for such analysis are shown in Table 3.42 and follow a problem I faced some time ago when I used to live at point B and needed to travel to point A (Figure 3.5).

As it turns out, I attempted all sorts of combinations to get to work. Naturally I cared for cost and travel time but also for comfort (I cannot handle more than 30 minutes in a bus, I guess you can picture the rest). Also, I was not very happy at expending a lot of time commuting. Each individual has different tastes and preferences, so just bear with the example for now (Figure 3.5).

Table 3.42 shows the summary of original criteria before the implementation of ranges. I have added the results of a dominance analysis and leave its proof to the reader. As you can see, alternatives 5 and 6 had a very similar travel time; I replaced the average (0.83) and used it for both.

Alternatives 6 and 11 had a very close cost (145 versus 151); I decided not to do any changes, after all $6 does buy me a meal.

Given I cannot handle bus travelling very well and I live in a place with cold temperatures (hence cannot walk for more than 20 minutes without freezing), I had established a comfort criterion that assigns five points or less to such undesirable choices.

A satisfaction criterion sorts choices and eliminates alternatives that fail to reach a minimum threshold level, and in this case, any choice with a comfort smaller than five is ruled off by the analysis as shown in Table 3.43. Again, each decision maker may have a different preference and this value is just an example.

Now let us make use of the lexicographic approach. Consider the case once satisfaction has been surpassed that is already used to remove those options that go beyond a certain threshold level.

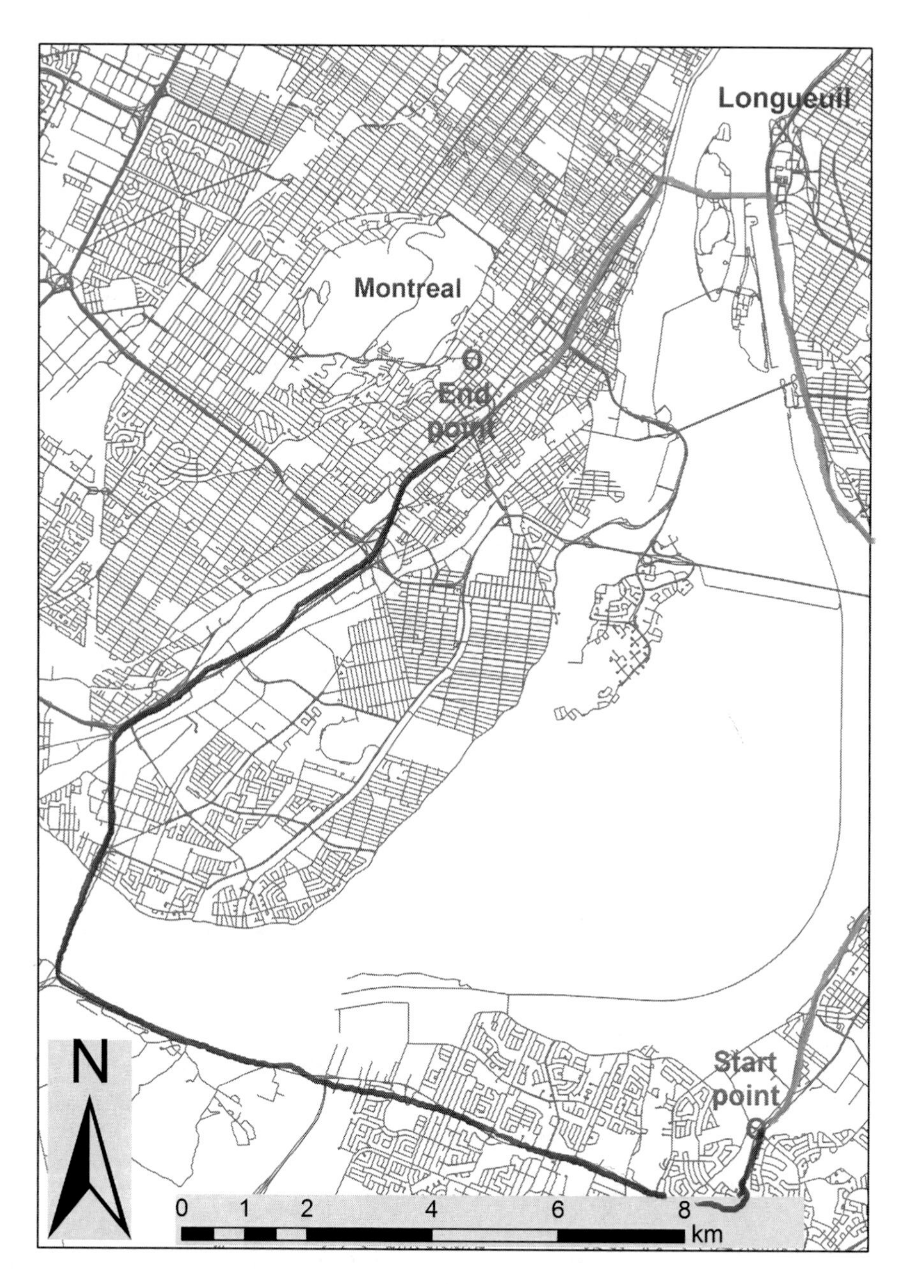

FIGURE 3.5
Inferiority with ranges. (Courtesy of Google Maps.)

TABLE 3.43

Applying a Satisfaction Rule

Option	Time (Hours)	Cost ($/Month)	Comfort	Satisfaction
4. Car	0.42	300	10	
7. Car–metro(2)	1	305	9	
6. Car–metro(1)	0.83	145	8	
8. Car–train(1)	0.92	175	7	
9. Car–train(2)	1	175	7	
10. Walk–train	1.25	125	6	
11. Car–bus	0.67	151	5	Fail satisfaction
2. Bike	1.18	0	4	Fail satisfaction
1. Walk	4.5	0	3	Fail satisfaction
3. Bus	1	101	1	Fail satisfaction
5. Walk–bus	0.83	101	0	Fail satisfaction

TABLE 3.44

Lexicographic Sorting: Cost Then Travel Time

Alternative	Time (Hours)	Cost($/Month)	Comfort	Preference
10. Walk–train	1.25	125	6	First
6. Car–metro(1)	0.83	145	8	Second
8. Car–train(1)	0.92	175	7	Third
9. Car–train(2)	1	175	7	Fourth
4. Car	0.42	300	10	Fifth
7. Car–metro(2)	1	305	9	Sixth

After comfort, I really care about cost, even more than travel time. Table 3.44 shows the final result of applying a lexicographic rule to sort based on travel time first and cost second.

As a result of applying the satisfaction and lexicographic criteria, it turns out that alternative 10 (walk then take the train) is the best choice. I did actually implement this solution for about 2 years.

3.10.2 Utility Approach

The last method we are going to cover is that of the utility approach. It consists of joining together all criteria under one sole indicator called utility. Several considerations must be taken. First, units and amounts will likely differ, hence a rescaling may be necessary.

Also, the sense of improvement may not be the same; therefore for those elements that provide us with decreasing values, we need to reverse their sense. The following equation is commonly applied to rescale any set of numbers by using their maximum (max), minimum (min) and every given value: $\left(1 - \frac{max-value}{max-min}\right) * newscale$.

Sometimes, there is a need to reverse the sense of a given amount to represent the opposite so that it matches the sense of improvement of all other amounts on the analysis. For these cases, the rescalation formula takes the following expression: $\left(\frac{max-value}{max-min}\right) * newscale$. Let us look at an example.

3.10.3 An Example of the Utility Approach

Imagine you are currently renting and spending $1000 a month (including transportation); you just realized that instead of wasting your money on rent you rather build some equity by purchasing a condo (in a way, a good portion of your monthly mortgage payment accumulates, and when you sell your property, you reap the benefits of it).

You prefer to have a place close to your workplace, or at least within an easy commute, and you wish to remain close to your current level of spending; the amount of space is also important (to fit your stuff).

Let us assume you really care for the cost and so you give it 50% of the overall priority, then you care about the comfort of your place and your travel (say 40%), and finally, you care about travel time (10%). Let us pretend we do not care about the downpayment. Table 3.45 shows a summary of the corresponding values for this problem.

Before we proceed to combine all amounts, we need to rescale columns 1 and 2 to a 0 to 10 scale using the following equation: $\left(\frac{max-value}{max-min}\right) * 10$. We use the average for travel time (however, notice an option with largest variability represents higher degree of uncertainty). Cost scale is reversed so that a 10 represents the cheapest and a 0 the most expensive. The time scale was also reversed. A longer commute time received less points on a 0–10 scale.

TABLE 3.45

Utility Approach: Summary of Available Information

Alternative	Monthly Cost	Travel Time	Comfort	Space	Downpayment
A	1100	30–40	6	7	5
B	1100	30–40	8	7	5
C	1350	10–15	9	4	15
D	1590	5–10	10	8	15
E	1350	10–15	9	8	15
F	1100	20–50	5	10	5
G	1100	10–15	9	8	15
H	1100	10–15	9	6	5

TABLE 3.46

Utility Approach: Re-Scaling

Alternative	Cost (Reversed Scale)	Time (Reversed Scale)	Comfort	Space
A	$\left(\frac{1590-1100}{1590-1100}\right)*10=10$	$\left(\frac{35-35}{35-7.5}\right)*10=0$	6	7
B	$\left(\frac{1590-1100}{1590-1100}\right)*10=10$	$\left(\frac{35-35}{35-7.5}\right)*10=0$	8	7
C	$\left(\frac{1590-1350}{1590-1100}\right)*10=4.9$	$\left(\frac{35-12.5}{35-7.5}\right)*10=8.2$	9	4
D	$\left(\frac{1590-1590}{1590-1100}\right)*10=0$	$\left(\frac{35-7.5}{35-7.5}\right)*10=10$	10	8
E	$\left(\frac{1590-1350}{1590-1100}\right)*10=4.9$	$\left(\frac{35-12.5}{35-7.5}\right)*10=8.2$	9	8
F	$\left(\frac{1590-1100}{1590-1100}\right)*10=10$	$\left(\frac{35-35}{35-7.5}\right)*10=0$	5	10
G	$\left(\frac{1590-1100}{1590-1100}\right)*10=10$	$\left(\frac{35-12.5}{35-7.5}\right)*10=8.2$	9	8
H	$\left(\frac{1590-1100}{1590-1100}\right)*10=10$	$\left(\frac{35-12.5}{35-7.5}\right)*10=8.2$	9	6

TABLE 3.47

Utility Approach (Relative Weights)

Alternative	Cost (0.5)	Time (0.1)	Comfort (0.2)	Space (0.2)	Utility
A	10	0	6	7	7.6
B	10	0	8	7	8
C	4.9	8.2	9	4	5.87
D	0	10	10	8	4.6
E	4.9	8.2	9	8	6.67
F	10	0	5	10	7
G	10	8.2	9	8	9.22
H	10	8.2	9	6	8.82

It is perhaps clear now (after seeing Table 3.45) that those options with the higher travel time and cost receive a zero in terms of utility (Table 3.46). Table 3.47 applies the utility weights and computes a final utility index that reveals a clear-cut answer to our problem of choosing a condo. Clearly, option G is the best alternative given the considered criteria.

Exercises

1. $MAX x_1 - x_2$
   $3x_1 + 4x_2 \leq 12$
   $4x_1 + 3x_2 \leq 12$
   $x_1, x_2 \geq 0$
   In this case, notice the objective is to maximize and it moves in the direction of x_1, so if we increase it, we improve it. So how far can x_1 go?

The answer is $x_1 = 4$ and $x_2 = 0$ because any value we take of x_2 will have a negative impact on the objective.

2. $MAX - x_1 + x_2$
   $3x_1 + 4x_2 \leq 12$
   $4x_1 + 3x_2 \leq 12$

   In this case, notice the objective is to maximize and it moves in the direction of x_2, so if we increase it, we improve it. So how far can x_2 go? The answer is $x_2 = 5$, but wait, nothing is preventing us from taking negative values of x_1. There is not a non-negativity constraint, so $x_1 = -1$ will help improve the objective, $x_1 = -100,000,000$ will help even more and $x_1 = -100,000,000,000,000,000$ will be even better. Now how far can we go? Problem is unbounded, and does not have a solution because the sense of improvement is in the direction where we lack a boundary.

3. $MINx_1 + x_2$
   $3x_1 + 4x_2 \geq 12$
   $4x_1 + 3x_2 \geq 12$
   $x_1 + x_2 \leq 10$

   Notice the objective is now to minimize, so we want to move in the x_1 and x_2 decreasing direction; again, there is not a non-negativity constraint (i.e. $x_1 \geq 0, x_2 \geq 0$), but the inequalities give us a minimum point, actually the very last constraint is not needed. So there will be an answer at the intersection of $3x_1 + 4x_2 \geq 12$ with $4x_1 + 3x_2 \geq 12$.

4. Let us revisit the municipal engineer in trouble. We modify the problem and assume the mayor had promised to accomplish at least 10% replacement rates per month, meaning each month you have to replace at least 1000 water meters (or more), which on a daily basis comes to be 50 m (assuming 20 business days per month). A skilled external worker can replace 5 m per day (100 per month), and a skilled internal worker can replace 4 m per day. This constraint looks like $4x_1 + 5x_2 \geq 50$. Let us re-estimate how many workers you should hire under this new circumstance. Table 3.48 sets the problem in the linear programming language.

5. Identify inferior and non-inferior alternatives, and for those inferior alternatives, indicate which alternative dominates them (Table 3.49).

TABLE 3.48

Problem Formulation for Example 1

Element	Composition	Units
Replacement	$4x_1 + 5x_2 \geq 50$	Water meter per day
Legislation	$-2x_1 + x_2 \leq 0$	Workers ratio per month = day
Minimize	$10x_1 + 12x_2$	Dollars per day

TABLE 3.49

Base Information

Alternative	x_1	x_2	$maxZ_1$	$maxZ_2$	$minZ_3$	Non-inferiority
A	2	0	6	−2	2	
B	4	1	10	−2	5	
C	6	5	8	4	11	
D	6	7	4	8	13	
E	3	6	−3	9	5	
F	1	5	−7	9	6	
G	0	3	−6	6	12	
H	0	1	−2	2	1	

Solutions

1. In this case, notice the objective is to maximize, and it moves in the direction of x_1, so if we increase it, we improve it. So how far can x_1 go? The answer is $x_1 = 4$ and $x_2 = 0$ because any value we take of x_2 will have a negative impact on the objective.
2. In this case, notice the objective is to maximize and it moves in the direction of x_2, so if we increase it, we improve it. So how far can x_2 go? The answer is $x_2 = 5$, but wait, nothing is preventing us from taking negative values of x_1; there is not a non-negativity constraint, so $x_1 = -1$ will help improve the objective, $x_1 = -100{,}000{,}000$ will help even more and $x_1 = -10{,}0000{,}00000{,}0000{,}000$ will be even better. Now how far can we go? The problem is unbounded and because the sense of improvement is in the same direction of lack of bound is that we cannot solve it.
3. Notice the objective is now to minimize, so we want to move in the x_1 and x_2 decreasing direction; again, there is not a non-negativity constraint (i.e. $x_1 \geq 0; x_2 \geq 0$), but the inequalities give us a minimum point. Actually, the very last constraint is not needed. So, there will be an answer at the intersection of $3x_1 + 4x_2 \geq 12$ with $4x_1 + 3x_2 \geq 12$.
4. The solution is $x_1 = 3.571429$ and $x_2 = 7.142857$.
5. Alternatives F and G are dominated by alternative E; the rest are non-inferior alternatives.

4

Probability and Statistics

4.1 Statistics

Often, the civil engineer is faced with data and the need to use it to justify decisions or make designs. It is common that the amount of data is large and the engineer requires simple and fast understanding of the phenomenon at hand, whether it is water consumption for a municipality, population growth, trips of passengers, vehicle flows, infrastructure condition, collision on highways, etc.

Statistics plays a key role in providing meaningful and simple understanding of the data. The most famous element is the mean; it represents the average value observed: in a sense that it is like having all observations condensed into one that is at the midpoint among all of them. The mean is obtained by simply summing over all observations and dividing by their number; sometimes, we calculate an average for a sample and not for the entire population:

$$\frac{\sum_{i=1}^{N} x_i}{N} = u.$$

One problem of having all elements represented by the average (or the mean) is that we lose sight of the amount of variation. To counteract this, we could use the variance which measures the average variation between the mean and each observation:

$$\frac{\sum_{i=1}^{N} (u - x_i)}{N} = v.$$

The problem with the variance is that it could potentially go to zero when observations above (positive values) and below (negative values) could possibly cancel each other. The way to solve this issue is by obtaining the standard deviation (*sd*), which takes the variance and squares the values: this is one of two possible ways to remove the positive and negative values. The other one is by using the absolute value. The advantage of the standard deviation is that its units and magnitude are directly comparable to those of the original problem, once you square and apply the square root, obtaining an indication of the average amount of variation on the observations (Figure 4.1).

$$\sqrt{\frac{\sum_{i=1}^{N}(u - x_i)^2}{N}} = std.\ dev.$$

In several instances, it is useful to bin the data in ranges and plot this information using a histogram. The histogram must be constructed in a way that it provides quick visual understanding.

There are two other measures: kurtosis and skewness. Kurtosis is a measure of how much the distribution is flat or peaked and skewness of whether the distribution is asymmetric towards the right or left of the mean.

Having data that follow a normal distribution is important because it allows you to use the normal distribution to standardize variables for estimation and prediction analysis.

The most common application of a normality plot is to identify outliers that scape the normal behaviour: this is easily identified by the enveloping lines. Once the spread of observations scape such enveloping lines, one would cut such items from the set of observations, which is typically refered to as cutting off the tails of the data (Figure 4.2).

The next tool that is commonly used are boxplots. They represent several elements together in a graphical way. The midpoint of a boxplot corresponds to the mean, and the edges of the box represent equal percentiles; hence, an unbalance is an indication of skewness. Outliers are also represented as dots

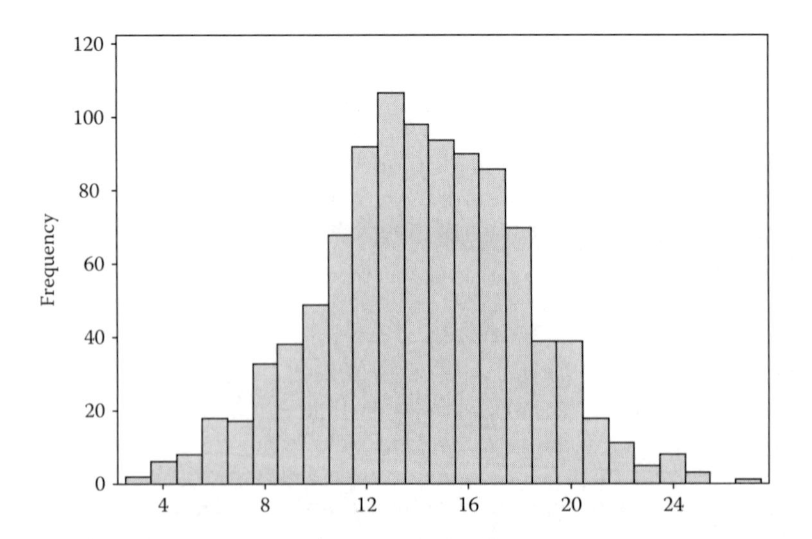

FIGURE 4.1
Sample histogram.

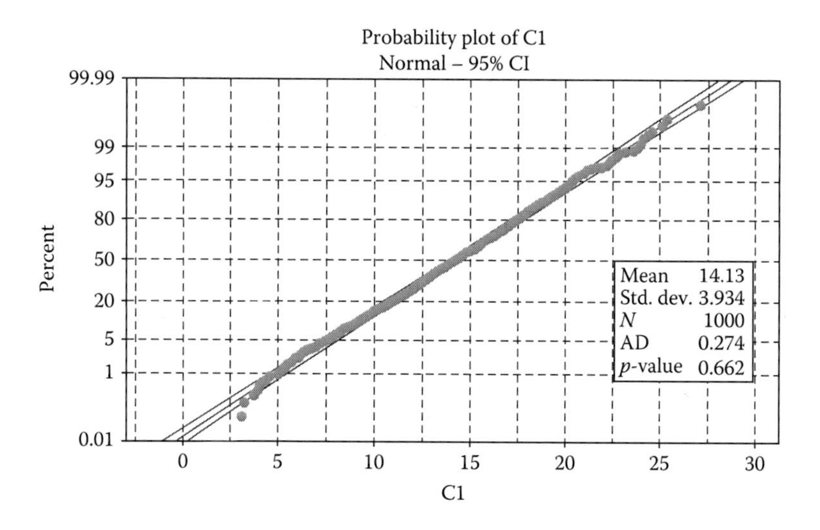

FIGURE 4.2
Normality test plot.

on the outer edge of each boxplot. Boxplots could be used to rapidly compare two data sets and conclude whether they are representative of the same population or not.

In the previous figure, one can see that both samples do not belong to the same population, the gross area does not match, the first sample has much less variation and the second sample not only has more variation but also many outlier values concentrated on the high values (between 25 and 30).

4.1.1 Statistics: A Simple Worked-Out Example

Suppose you have observed collisions and several possible causal factors related to vehicle collisions. For simplicity, we will have a very small database of four observations; in real life, you will have hundreds – if not thousands – of observations. Our four elements are the number of collisions: the speed of the vehicle at the time of the collision, the length of the road segment and the density of intersections (per kilometre) (Table 4.1 and Figure 4.3).

TABLE 4.1

Sample Data Deterioration

y_{obs}	v (km/h)	L (km)	I (/km)
9	100	1	0.5
7	70	1	1
11	110	1	0.25
7	60	1	2

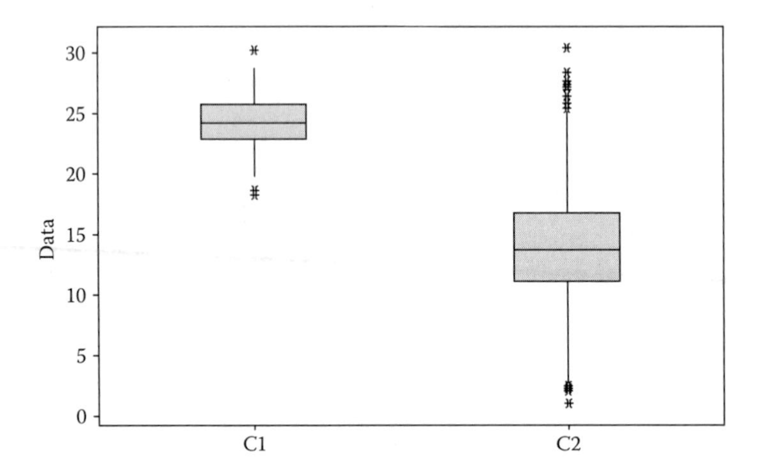

FIGURE 4.3
Boxplot of C1 and C2.

In the case of length, the mean is equal to the value of each and every observation, and the standard variation is zero. A very rare scenario for real-life problems. However, as you will see, if this is the case and you are trying to estimate explanatory contribution of factors, you can simply ignore the length from your calculations.

For the number of collisions, we have an average of $((9+7+11+7)/4) = 8.5$ collisions. The variance for this example and the standard deviation could be estimated for any amount, even the responses y_{obs} for speed v (km/h) are estimated in Table 4.2.

As you can see, the use of the variance poses a problem, because it does not remove the negative signs; in this particular example, we observe a critical case in which the variance goes to zero, which could be interpreted as having no dispersion (variability) on the observed values of speed. This of course is not true. We can observe values as little as 60 and as high as 110.

The standard deviation correctly captures the dispersion in the same units as the original factor (km/h). Our standard deviation tells us that average

TABLE 4.2
Variance and Standard Deviation (Std. Dev.)

y_{obs}	v (km/h)	L (km)	I (/km)	Variance	Std. Dev.
9	100	1	0.5	15	225
7	70	1	1	−15	225
11	110	1	0.25	25	625
7	60	1	2	−25	625
Summation=	85			0	20.62

variability or dispersion on speed is 20.62 km/h. If you combine this with the mean, you end up with a good grasp of what is going on; observed speed on average fluctuates from 64.38 (mean − std. dev. = 85 − 20.62) to 105.62 (mean + std. dev.).

As a matter of fact, if the values of speed follow a normal distribution, then there is a rule that can be implemented to measure with 95% confidence the dispersion using the mean and the standard deviation. The rule simply states that you add to the mean 1.96 standard deviations (for fast calculation, many authors round this to two times the standard deviation). Then you repeat the same but you subtract 1.96 standard deviations from the mean. These two values capture the spread of 95% of the data and we will use it later on in this book.

4.1.2 Statistics: A More Realistic Example

Let us take as an example the data for sanitary pipes (Table 4.3). This example will be useful for learning how to construct a histogram from scratch. I will also provide you with the sequence of steps to do this automatically using Excel. The reader should note that at the time I wrote this book, the latest version of Excel was 2013; in the future, the location of bottoms could change, but I am confident the sequence of steps will remain the same.

Let us produce a histogram. I will do it manually only this time, so that you understand what to get when you do it automatically. Let us do it for the rim elevation. This elevation tells the engineer the absolute value of elevation of the bottom of the manhole and is important when you need to do hydraulic calculations related to the flow of polluted water, passing-by manholes. Remember that manholes are used to change the direction of the flow, connecting two pipes. The slope could change as well.

First, we find the smallest (680.2) and the largest (685.9) values, and then you decide how many bins you wish to use. In my case, I decided to use five bins (I recommend the reader to repeat the calculations with seven bins); it is always advisable to have an impair number of bins wherever possible. Now let us find out the range for the bins: we take the span between smallest and largest (685.9 − 680.2 = 5.7) and divide it by the number of bins ((5): 5.7/5 = 1.14). This latter amount is called the increment. Now take the smallest value and add the increment (680.2 + 1.14 = 681.34); this creates the first bin interval from 680.2 to 681.34. We take the high threshold of the previous interval and add the increment to obtain the next bin values 681.34 + 1.14 = 682.48, similarly the third bin goes up to 683.62 and the following bin to 684.76 and the last bin to 685.9. Now, we simply count how many items we observe per bin. The size of each bin will reflect the frequency of observations (count) that belong to it. Table 4.4 gives you a tally of values per bin.

With the counts and the values of intervals for the bins, we can proceed to create our histogram which is presented in Figure 4.4.

TABLE 4.3
Manhole Data

Source	Year of Construction	Rim Elevation	$(u - x_i)^2$	
a	1950	684.39	3.2	
a	1951	684.2	2.55	
a	1952	685.9	10.87	
a	1953	682.45	0.02	
a	1953	682.31	0.09	
a	1954	681.05	2.41	
a	1954	680.77	3.36	
a	1954	680.56	4.17	
a	1955	680.2	5.77	
a	1955	680.42	4.76	
a	1956	682.12	0.23	
b	1957	682.07	0.28	
b	1958	681.42	1.40	
b	1958	682.13	0.22	
b	1959	683.90	1.68	
b	1959	683.70	1.20	
b	1959	683.6	1.00	
b	1960	684.10	2.24	
b	1960	684.15	2.40	
b	1961	682.36	0.06	
b	1962	682.85	0.06	
	Mean=	682.60	1.55	=Std. dev.

TABLE 4.4
Manhole Data

Low Bound	High Bound	Count
680.2	681.34	5
681.34	682.48	7
682.48	683.62	5
683.62	684.76	3
684.76	685.9	1

If you want to do it in Excel automatically, you need first to install the add-in called *Data Analysis*. Follow this procedure:

1. Go to the *FILE* tab.
2. Select *Options* located at the very bottom of the left ribbon, and a new window will open.

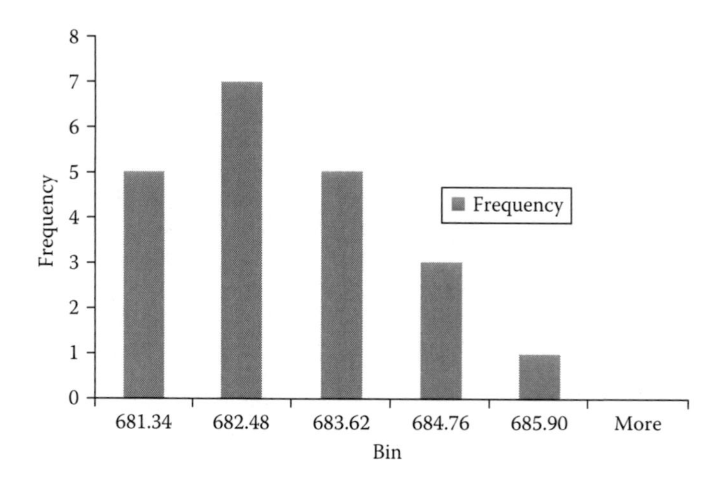

FIGURE 4.4
Histogram.

3. Select *Add-Ins*.
4. Centred at the bottom you will see *Manage: Excel Add-ins*; hit on the *Go...* next to it.
5. A new window will open; you need to click a check next to *Analysis ToolPak*. I would also advise you to do the same for *Solver Add-in* as we will use later.
6. Hit *OK*.

The new add-ins can be found at the *DATA* tab. Click on *Data Analysis*, select *Histogram*, specify the *Input range* that contains the observations for which the histogram will be constructed, specify the range (you must have written this on Excel's cells in column format) and hit *OK*; you will obtain a figure similar to that previously presented.

Let us concentrate our attention now on the construction of boxplots. Let us select the *year* as our variable of interest. Let us also presume that our data came from two separate sources (this has been identified in Table 4.3 with the name source on the first column). If you estimate the median (in Excel, use the command *Median(range)*) for the first and third quartiles (*QUARTILE(range,1) and QUARTILE(range,3)* along with the minimum value (*MIN(range)*) and maximum values (*MAX(range)*), then you would obtain all the information you need to produce boxplots. For our previous example and for source a and b, our calculations show the values shown in Table 4.5.

Producing a graph from Excel is not so straightforward; however, it is possible. Compute the distances between the mean and the third and

TABLE 4.5

Boxplots

Element	Source a	Source b
Maximum	1956	1962
Quartile 3	1954.5	1960
Median	1954	1959
Quartile 1	1952.5	1958.25
Minimum	1950	1957

TABLE 4.6

Boxplot Elements for Graph

Element	Source a	Source b
Level	1952.5	1958.25
Bottom	1.5	0.75
Top	0.50	1
Whisker top	1.5	2
Whisker bottom	2.5	1.25

first quartiles. This number will define the height of the two bars stacked on the boxplot. The first quartile will be the base reference level to plot each group and will help to compare them. The other element you need to compute is the whiskers; for this, subtract the max from the third quartile and the first quartile from the minimum. These calculations are shown in Table 4.6.

To do this in Excel, create Table 4.6 on your spreadsheet and then follow this procedure.

1. Select the level, bottom and top for both source a and source b.
2. Hit *INSERT*.
3. Hit *Column Chart*.
4. Hit *stacked column*.

   Figure 4.5 illustrates steps 1–4.
5. Select *change row by column* (Figure 4.6).
6. Click on the graph and select the top area of each bar.
7. Click on *Add chart element* and then select error lines.
8. Click on *more error bars options*.

   Figure 4.6 illustrates steps 5–8.

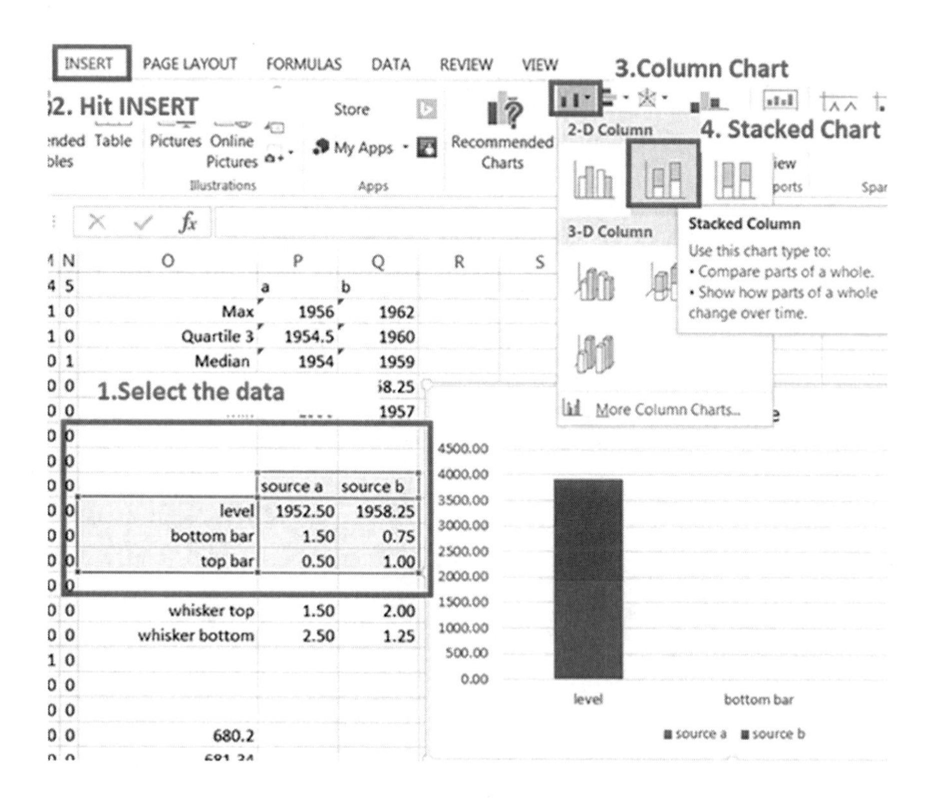

FIGURE 4.5
Steps 1–4.

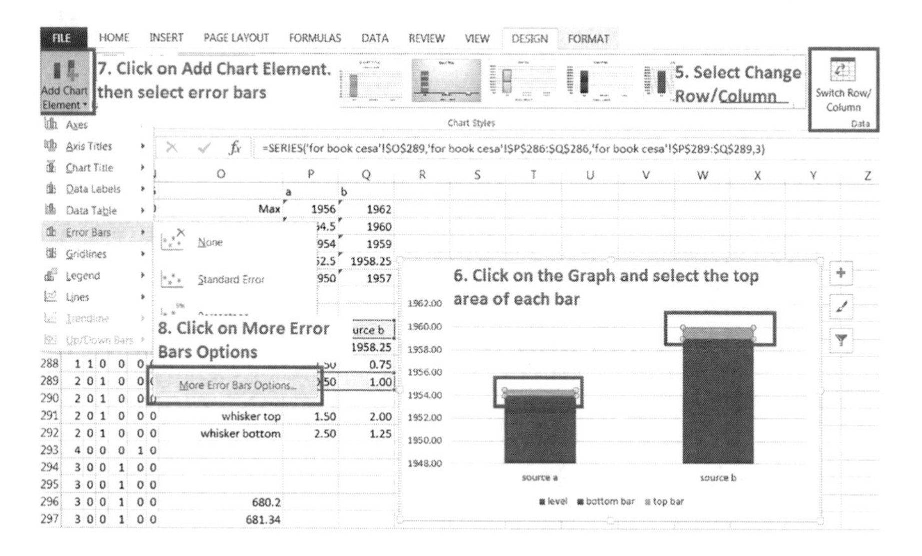

FIGURE 4.6
Steps 5–8.

9. Select *Plus* from the right-side ribbon.
10. Select *Custom* from the right-side ribbon.
11. Select the Top Whisker to be the positive error value.

    Figure 4.7 shows steps 9–11.
12. Repeat steps 7–11 for the bottom whisker. However, in step 9 choose *Minus*, and in step 11, select the Bottom Whisker to be the negative error value.

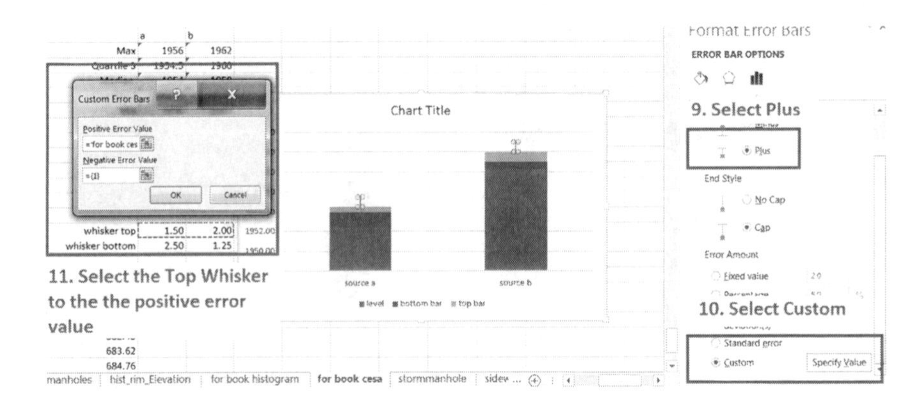

FIGURE 4.7
Steps 9–11.

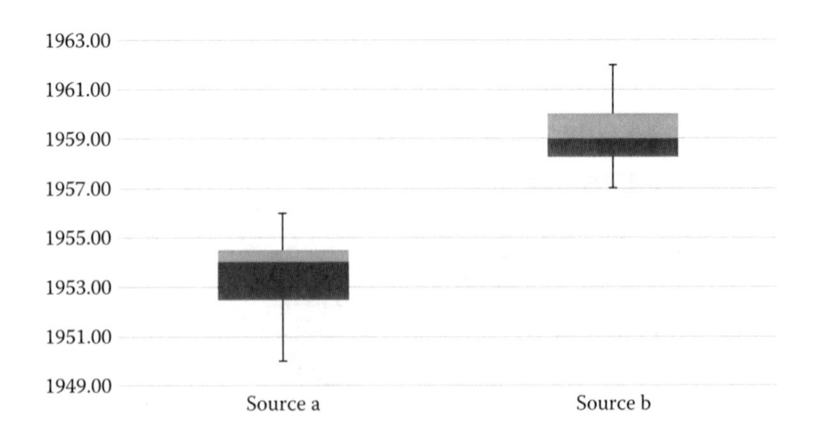

FIGURE 4.8
Final boxplots.

13. Choose again from the bar graph the bottom portion of the bar, click on the top tab called *FORMAT* and select *Shape Fill* to *No fill*. The final boxplots are produced after this. I have shown them in Figure 4.8.

Notice from this you can easily interpret that the two samples do not belong to the same population; it is clear from the comparison of both boxplots that they do not overlap in any way.

4.2 Probability

Probability refers to the likelihood of an event. The typical notation used for probability of an event is the letter P followed by a term inside a parenthesis that provides information regarding the elements whose probability is calculated, and whether such probability is dependent on other elements. Per instance $P(x)$ reads as the probability associated to an event x; when such event is discrete (takes on a series of finite values), one can simply divide the number of times that $x = a$ is observed by the total number of observations.

For example, if we are interested in the likelihood of observing the density of intersections I equal to 0.5, we can easily see that this happens in one out of the four observations and hence, the probability is $1/4 = 0.25$. The likelihood of observing an event can be conditioned to observing values bigger or equal to a certain level; for instance, if the legal speed was 90 km/h, what would be the probability of observing values beyond the speed limit ($P(v \geq 90)$)? Such values would be the count of observations above the speed limit (2) over the total number of observations (4) so $P(v \geq 90) = 0.5$ or 50%.

Typically, the observations that we have were collected beforehand, they represent the results from some experiments and their values could be conditioned by other elements that we can or cannot observe. For instance, the speed v could be conditioned by the age of the driver, younger drivers may tend to overspeed while older drivers would drive below the speed limit and adult drivers will likely drive around the legal speed; we could express a probability of driving above the legal speed conditioned on driver's age a, and this would be expressed as $P(v \geq 90/a)$.

We would use conditional probabilities also when we want to know the likelihood of observing an outcome from a specific type of event. The civil engineer will face often the need to use machines to estimate properties of a structure or the soil which cannot be observable directly. The corrosion of re-bars inside a column, the strength of a pavement structure and the bearing capacity of the soil are some examples you will face in your professional life.

For instance, consider the case in which you have a machine that allows you to estimate the strength of a pavement structure. For instance, think of a *deflectometer* that reads the amount of deflection on the pavement when a given weight hits it. This aim of using this machine is to characterize the strength of the pavement structure, which is something that we cannot observe directly. The problem is that the machine has a rebound and sometimes the rebound bias the readings, so often the reading indicates a given value but the true value does not perfectly match. Consider, for instance, where we use this machine 100 times, and the results of how many times the machine identifies the correct level of strength are given in Figure 4.9.

For pavement with truly low bearing capacity, the machine has successfully identified the presence of a weak pavement 26 times, but mistakenly indicated a fair pavement strength on five occasions and 1 time a strong pavement. Similar results could be interpreted for medium and high pavement strength. The probabilities associated with the state of nature can be measured using the subtotals given at low, medium and high pavement strength as shown in Figure 4.10. Similarly, we could obtain marginal probabilities for the experiment with the *deflectometer*, which would tell you that the *deflectometer* detects 39% of the time a weak soil, 35% of the time a fair pavement and 26% of the time a strong pavement.

The problem is that we want to know how often the machine is right within each category. To solve this we use the marginal probabilities of the experiment to condition the probability values of each observation previously obtained. Figure 4.11 shows the calculations and results. These values can be expressed as conditional probabilities of observing each real phenomenon result given that the experiment is detecting a particular state.

The values obtained in Figure 4.11 reveal that 67% of the time the experiment correctly identifies a weak pavement, 20% of the time incorrectly labelled it as a medium (being really weak) and 13% of the time incorrectly labelled it as high, but actually it was weak. Similarly, for a pavement whose real strength lies in the medium range, 14% of the time the experiment results

		Experiment result: deflectometer result			
		Weak	Fair	Strong	Subtotals
Real state of nature: pavement strength	Low	26	5	1	32
	Medium	8	21	7	36
	High	5	9	18	32
Subtotals		39	35	26	100

FIGURE 4.9
Conditional probabilities.

		Experiment result: deflectometer result				Marginal probability of the state of nature
		Weak	Fair	Strong	Subtotals	
Real state of nature: pavement strength	Low	26	5	1	32	0.32
	Medium	8	21	7	36	0.36
	High	5	9	18	32	0.32
Subtotals		39	35	26		
Marginal probability of the experimental result		0.39	0.35	0.26		

FIGURE 4.10
Marginal probabilities.

		Experiment result: deflectometer result		
		Weak	Fair	Strong
Real state of nature: pavement strength	Low	0.26/0.39 =0.67	0.05/0.35 =0.14	0.01/0.26 =0.04
	Medium	0.08/0.39 =0.20	0.21/0.35 =0.60	0.07/0.26 =0.27
	High	0.05/0.39 =0.13	0.09/0.35 =0.26	0.18/0.26 =0.69
Subtotals		1	1	1
Marginal probability of the experimental result		0.39	0.35	0.26

FIGURE 4.11
Conditional probabilities.

in a wrong identification as low, 60% of the time correctly identifies it and 26% of the time the experiment labelled the pavement as high strength, but actually it was medium. A similar interpretation can be done for the experiment in its ability to identify a strong pavement. As you can see, the totals now add to 100% (or one in decimal base calculations).

4.3 Linear Regression

This is a very common technique to find the relationship between a set of elements (independent variables) and a response (dependent variable). This technique is called linear regression and should not be confused with the ordinary least squares (presented in Chapter 5).

The idea behind linear regression is to find out a set of factors that have explanatory power over a given response. Think first of one factor alone, if every time such factor moves so does the response (and in the same direction), then the factor and the response are (positively) correlated and the factor could be though as influencing the response. Now presume the same factor moves but the response does not; in this case, it will be very difficult to give some explanatory power to such factor.

The problem becomes more challenging when we add more factors because all factors could potentially be influencing the response. Imagine you have a large database with several factors (consider per instance the one presented in Table 4.7) but some of them exhibit a repetitive nature.

TABLE 4.7

Sample Data Deterioration

y = Response	x_1	x_2	x_3
0.18	60,000	3.96	0.04
0.45	122,000	3.96	0.06
0.67	186,000	3.96	0.08
0.54	251,000	3.96	0.04
0.95	317,000	3.96	0.06
1.08	384,000	3.96	0.08
0.83	454,000	3.96	0.04
1.46	524,000	3.96	0.06
1.48	596,000	3.96	0.08
1.17	669,000	3.96	0.04
1.89	743,000	3.96	0.06
1.96	819,000	3.96	0.08
0.88	897,000	3.96	0.02
1.82	976,000	3.96	0.04
1.67	1,057,000	3.96	0.05
2.66	1,139,000	3.96	0.07
1.30	1,223,000	3.96	0.02
1.43	1,309,000	3.96	0.03
2.40	1,396,000	3.96	0.05
2.39	1,486,000	3.96	0.04

TABLE 4.8

Sample Data Deterioration

y = Response	x_1	x_2	x_3
0.18	60,000	3.96	0.04
0.54	251,000	3.96	0.04
0.83	454,000	3.96	0.04
1.17	669,000	3.96	0.04
1.82	976,000	3.96	0.04
2.39	1,486,000	3.96	0.04

Hence, estimating the influence of a given factor (say x_1) on the response would be easy if we select from the database a subset of observations where all other factors ($x_2, x_3, \ldots, x_n$) have constant values. Per instance from Table 4.7 you can see that x_2 is already constant at 3.96, we need somehow to held constant x_3, if you look carefully you will discover a pattern of 0.04, 0.06 and 0.08, the pattern eventually breaks. Let us select from this data set a subset containing only those observations where $x_3 = 0.04$. This subset is shown in Table 4.8.

At this stage, we have isolated changing values of x_1 only while all other factors are being held constant, out of this we can estimate the explanatory power of x_1 on y. In a similar fashion, we could take on values of $x_3 = 0.06$, $x_3 = 0.08$, etc. The problem with this is that if we are obtaining isolated measures of explanatory power for specific ranges of the other variables, how do we combine them?

Linear regression follows a similar principle without the issue of creating isolated data sets. It estimates the explanatory power of one variable on a given response by looking into how much this variable influences movement on the response. For this, it held all other variables constant, just as we did but without creating isolated groups.

Exercises

1. Obtain the mean and standard deviation for the response y of Table 4.7. Then prepare a histogram for the response.
2. Obtain the mean and standard deviation for the factor x_1 of Table 4.7. Then imagine the first 10 elements belong to source a and the second half of the elements (from 11 to 20) belong to source b; prepare a boxplot for them.
3. A pressure disk can be used to learn soil-bearing capacity. Presume you have used in the past year this device and the reliability

TABLE 4.9

Measuring Soil-Bearing Capacity

Real/Measured	Low	Medium	High
Low	60	36	4
Medium	25	96	5
High	5	6	114

of the device is summarized by the numbers presented in Table 4.9.

Estimate the probability of correctly predicting low, medium and high capacity. Hint: Use the marginal probability of the experimental result, and notice the summation of observations does not add to 100.

4. Linear regression. Estimate if there is any explanatory power for the year of construction on rim elevation (use the information contained in Table 4.3).
5. Linear regression. Estimate if there is any explanatory power for the year of construction on the bottom elevation of manholes for the data presented in Figure 4.12.
6. Now prepare boxplots for source a and b for the variable bottom of the manhole elevation. Estimate if the data came from the same population. (Answer: No it did not.)
7. Table 4.10 shows several variables associated to alternatives:
   a. Estimate mean and standard deviation for each variable (except the alternative).
   b. Obtain a histogram for each variable (except the alternative).
   c. Test normality on each variable (except the alternative).

Solutions

1. The following table shows the calculations for the standard deviation for the response y of Table 4.11.

   Figure 4.13 shows the histogram; the values of the bins can be changed according to specific needs.
2. Figure 4.14 shows the steps and final results of the estimation of the mean and standard deviation for factor x_1 in Table 4.7.

   Figure 4.15 shows the boxplots for both samples. As you can see, they do not overlap in anyway and therefore are deemed as belonging from two different populations.

   It is important to indicate that these numbers came from one single exercise, but the nature of the observations is not the same; this is why neither the average nor the variation around the median matches.

DIAMETER__	YEAR	TROUGH_ELE
585.00000000000000	1985	678.07000000000000
585.00000000000000	1985	677.61000000000000
585.00000000000000	1985	677.73000000000000
585.00000000000000	1967	687.96400000000000
585.00000000000000	1968	684.21199999999900
585.00000000000000	1984	682.38999999999900
585.00000000000000	1984	681.40999999999900
585.00000000000000	1984	680.25000000000000
585.00000000000000	1984	679.95000000000000
585.00000000000000	1983	676.09000000000000
585.00000000000000	1985	679.94000000000000
585.00000000000000	1985	683.09000000000000
585.00000000000000	1985	682.55999999999900
585.00000000000000	1985	680.79999999999900
585.00000000000000	1985	681.21000000000000
585.00000000000000	1985	681.62000000000000
585.00000000000000	1985	679.39999999999900
585.00000000000000	1985	680.00000000000000
585.00000000000000	1981	677.76999999999900
585.00000000000000	1985	677.71000000000000
585.00000000000000	1981	677.02999999999900

FIGURE 4.12
Exercise 4.5.

TABLE 4.10
Choice of Commute to Work

Alternative	Time (Hours)	Cost ($/Month)	Comfort	Utility
1. Walk	4.5	10	3	5
2. Bike	1.18	10	4	6
3. Bus	1	101	1	5
4. Car	0.42	300	10	7
5. Walk–bus	0.82	101	0	6
6. Car–metro(1)	0.84	145	8	9
7. Car–metro(2)	1	305	9	7
8. Car–train(1)	0.92	175	7	6
9. Car–train(2)	1	175	7	5
10. Walk–train	1.25	125	6	4
11. Car–bus	0.67	151	5	5

TABLE 4.11

Solution for Exercise 4.1

y = **Response**	x_1	x_2	x_3
0.18	1.3605 – 0.18	1.1805	1.3935
0.45	1.3605 – 0.45	0.9105	0.829
0.67	1.3605 – 0.67	0.6905	0.4768
0.54	1.3605 – 0.54	0.8205	0.6732
0.95	1.3605 – 0.95	0.4105	0.1685
1.08	1.3605 – 1.08	0.2805	0.0787
0.83	1.3605 – 0.83	0.5305	0.2814
1.46	1.3605 – 1.46	−0.0995	0.0099
1.48	1.3605 – 1.48	−0.1195	0.0143
1.17	1.3605 – 1.17	0.1905	0.0363
1.89	1.3605 – 1.89	−0.5295	0.2804
1.96	1.3605 – 1.96	−0.5995	0.3594
0.88	1.3605 – 0.88	0.4805	0.2309
1.82	1.3605 – 1.82	−0.4595	0.2111
1.67	1.3605 – 1.67	−0.3095	0.0958
2.66	1.3605 – 2.66	−1.2995	1.6887
1.30	1.3605 – 1.30	0.0605	0.0037
1.43	1.3605 – 1.43	0.0695	0.0048
2.40	1.3605 – 2.40	−1.0395	1.0805
2.39	1.3605 – 2.39	−1.0295	1.0599
1.3605	=Total	Total=	8.7459
		/20=	0.4373
		Std. dev.=	0.6613

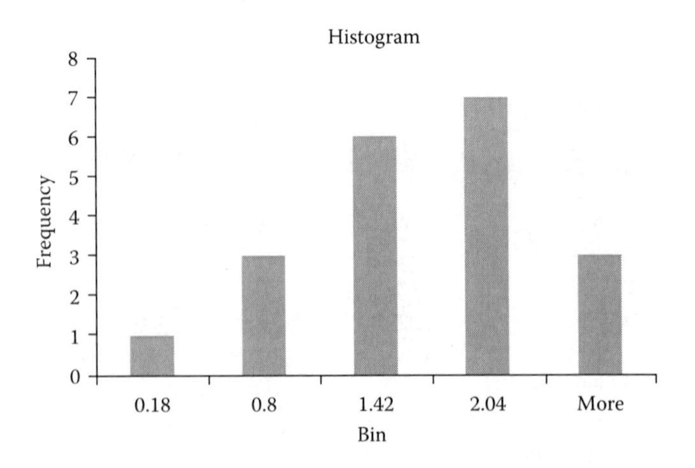

FIGURE 4.13
Solution Exercise 4.6.

	A	B	C
1		x_1	$(u - x_1)^2$
2		60,000	449,436,160,000
3		122,000	370,150,560,000
4		186,000	296,371,360,000
5		251,000	229,824,360,000
6		317,000	170,899,560,000
7		384,000	119,992,960,000
8		454,000	76,396,960,000
9		524,000	42,600,960,000
10		596,000	18,063,360,000
11		669,000	3,769,960,000
12		743,000	158,760,000
13		819,000	7,849,960,000
14		897,000	27,755,560,000
15		976,000	60,319,360,000
16		1,057,000	106,667,560,000
17		1,139,000	166,953,960,000
18		1,223,000	242,654,760,000
19		1,309,000	334,777,960,000
20		1,396,000	443,023,360,000
21		1,486,000	570,931,360,000
22	mean	730,400	
23		Sumation/2C	186,929,940,000
24		Std Dev	432353.9522

FIGURE 4.14
Solution Exercise 4.2 part (a) standard deviation.

3. Figure 4.16 shows the marginal and conditional probabilities.
4. Figure 4.17 shows the regression coefficient and *p*-value for year of construction on elevation. As seen, there is a positive explanatory power: the higher the value in year of construction, the higher is the rim elevation.
5. Figure 4.18 shows the regression coefficient and *p*-value for year of construction on bottom of manhole elevation. As seen, there is a positive explanatory power, the coefficient is 0.2476 with a *p*-value of 0.000. Hence, the higher the value in year of construction, the higher is the value of bottom of manhole elevation.

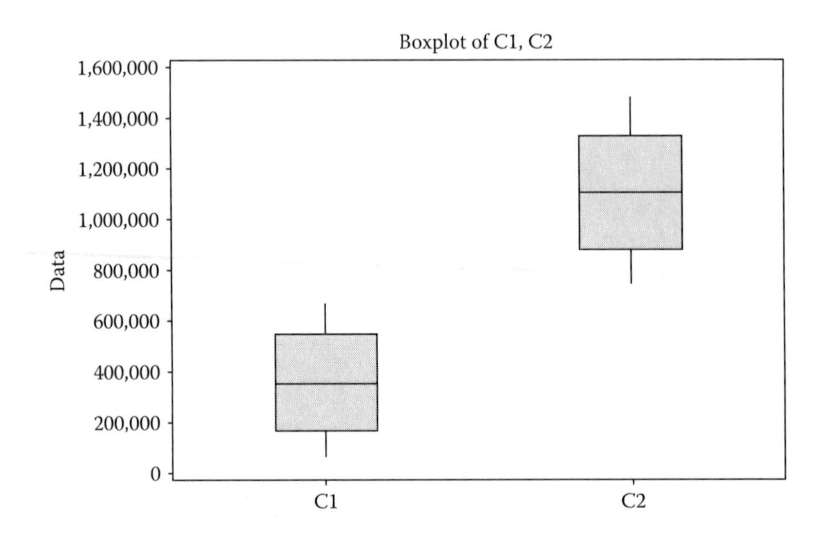

FIGURE 4.15
Solution Exercise 4.2 part (b) boxplots.

	A	B	C	D	E	F	G
1			Experiment result: deflectometer result				Marginal Probability of the state of Nature
2			weak	fair	strong	Subtotals	
3	Real State of Nature: Pavement Strenght	low	60.00	36.00	4.00	100.00	0.28
4		medium	25.00	96.00	5.00	126.00	0.36
5		high	5.00	6.00	114.00	125.00	0.36
6	Subtotals		90.00	138.00	123.00	351.00	
7	Marginal Probability of the experimental result		0.26	0.39	0.35		

FIGURE 4.16
Solution of Exercise 4.3 marginal and conditional probabilities.

```
. regress elevation year

      Source |       SS       df       MS              Number of obs =      21
-------------+------------------------------           F(  1,    19) =   54.90
       Model |  14.5498197     1  14.5498197           Prob > F      =  0.0000
    Residual |  5.0352461     19  .265012953           R-squared     =  0.7429
-------------+------------------------------           Adj R-squared =  0.7294
       Total |  19.5850658    20  .979253289           Root MSE      =  .51479

------------------------------------------------------------------------------
   elevation |      Coef.   Std. Err.      t    P>|t|     [95% Conf. Interval]
-------------+----------------------------------------------------------------
        year |    .247649   .0334227     7.41   0.000     .1776945    .3176035
       _cons |   200.8333   65.38123     3.07   0.006     63.98878    337.6778
------------------------------------------------------------------------------
```

FIGURE 4.17
Solution Exercise 4.4 regression: year of construction on rim elevation.

```
. regress var3 var2

      Source |       SS       df       MS              Number of obs =      21
-------------+------------------------------           F(  1,    19) =   10.70
       Model |  56.4996176     1  56.4996176           Prob > F      =  0.0040
    Residual |  100.334984    19  5.28078861           R-squared     =  0.3602
-------------+------------------------------           Adj R-squared =  0.3266
       Total |  156.834601    20  7.84173006           Root MSE      =   2.298

------------------------------------------------------------------------------
        var3 |      Coef.   Std. Err.      t    P>|t|     [95% Conf. Interval]
-------------+----------------------------------------------------------------
        var2 |   -.323864   .0990123    -3.27   0.004    -.5310992   -.1166288
       _cons |   1322.438   196.3091     6.74   0.000     911.5588    1733.318
------------------------------------------------------------------------------
```

FIGURE 4.18
Solution Exercise 4.4 regression: year of construction on bottom elevation.

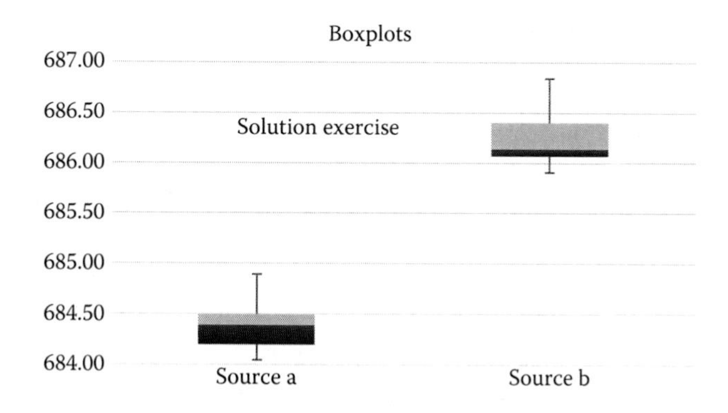

FIGURE 4.19
Solution Exercise 4.6.

Alternative	Time	Cost	Comfort	Utility
1	4.5	10	3	5
2	1.18	10	4	6
3	1	101	1	5
4	0.42	300	10	7
5	0.82	101	0	6
6	0.84	145	8	9
7	1	305	9	7
8	0.92	175	7	6
9	1	175	7	5
10	1.25	125	6	4
11	0.67	151	5	5
Mean	1.236364	145.2727	5.454545	5.909091
St.Dev.	1.054861	91.54342	3.055952	1.31111

FIGURE 4.20
Solution Exercise 4.7 part (a).

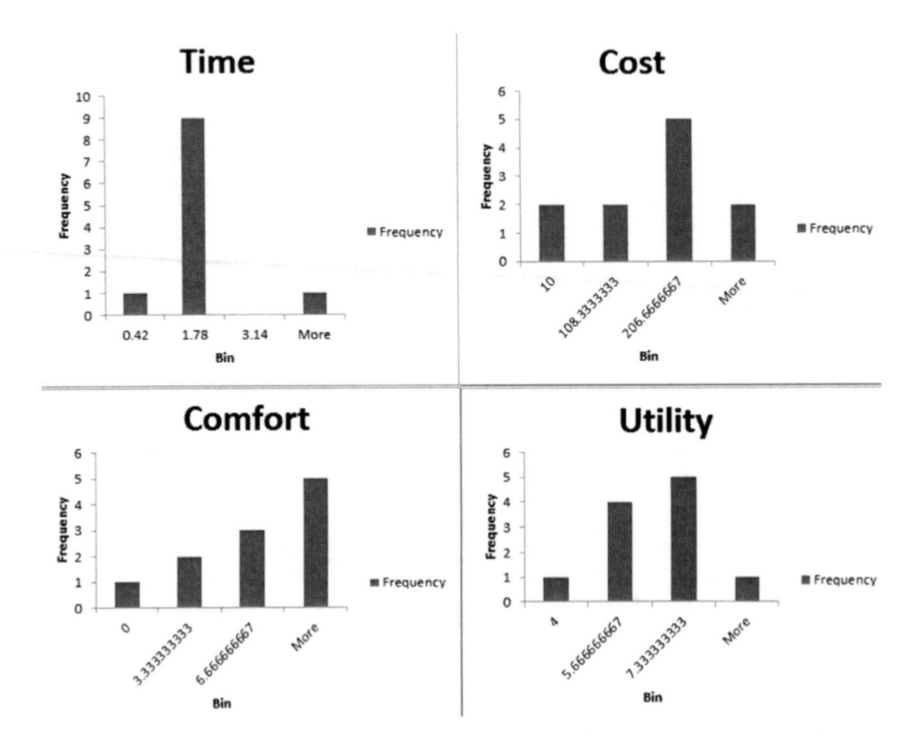

FIGURE 4.21
Solution Exercise 4.7 part (b).

6. Figure 4.19 shows the solution for Exercise 4.6.

   As you can see, the two samples do not come from the same population: the sample of source b has more variability above the median, and the sample from source a has a more variability above the median.

7. Figure 4.20 shows the solution for parts a and b.

   The histograms for time, cost comfort and utility are shown in Figure 4.21.

5

Estimation and Prediction

5.1 Introduction

This chapter goes beyond the basic concepts learned in Chapter 4 and uses them on practical applications. In this chapter, we learn two important concepts: estimation and prediction. As we will see, they are both interrelated. Estimation is the science of calibrating an equation that provides a representation of the real world containing only those elements deemed important given their capability to capture the phenomena being modelled. Prediction is the art of forecasting future states or values of a phenomenon. This typically has the purpose of making decisions to change the future. Prediction typically uses calibrated systems (estimation) along with some additional factors.

5.2 Estimation

Estimation is the ability to encapsulate a phenomenon mathematically. To be able to estimate, we need to create a functional form with at least one equation capable of capturing the nature of the phenomenon using measurable (observable) responses and factors that contribute to explain it. This sounds a lot like a regression analysis, and in many cases researchers could be tempted to run a regression and use it for estimation. However, you must note that a regression does not have the ability to capture the nature of the phenomenon in the relationships between the factors and the response. You could, however, prepare a linear regression to gain quick understanding of the relationship between observable or measurable factors and a response. However, you need to move ahead and estimate a mathematical representation of the factors and their contribution to the response. For this, we need to follow three steps:

1. Identify what elements play an important role on it.
2. Identify the functional form of the relationship between the response(s) and the factors.
3. Calibrate the mathematical expression.

Following these three steps will lead you to identify a functional form that contains a good representation of the nature of the phenomenon and the factors that play a role. You will use coefficients to calibrate the role of the factors and to adjust the mechanism(s) to match real-life observations.

After step 2, that is, when you obtain a representation of the system, you proceed to step 3 to calibrate the coefficients. This is however a tricky one. When the functional form is only one equation and you need to estimate several coefficients (unknowns) with an observed response per combination of factors, your only choice is to run a regression to estimate the values of the coefficients.

When you have a system of equations that predict the same number of indicators as unknown coefficients, then you could follow an iterative process to estimate them. In this case, it is typical that you would have to develop a strategy to calibrate the values in order to replicate those indicators (responses) to match their observed values.

Typically, you may need to iterate the value of the variables until you are able to replicate observed responses. For the remaining of this section, we concentrate our attention to an example.

5.2.1 Worked Example for Pavement Deterioration

This section describes a worked example that estimates pavement deterioration in the form of international roughness index (IRI). This index measures the total amount of vertical variation of a dumping system of a vehicle rolling through a road. The smoother the ride, the less vertical oscillations the system will experience. Vertical variations are measured from rest, and they all accumulate whether the tire moves up or down; it is like using the absolute value of the deviations.

The response for this example is an indication of the condition of the pavement, and a brand new pavement should be equivalent to a surface that has zero imperfections; hence, the ride should be very smooth and the value of IRI should be close to zero. As time passes, the layers underneath the pavement suffer from the effects of the weather and lose strength; this leads to loss of support which translates into deformations of the surface.

Thicker pavements are commonly associated to higher strength, however this must be contrasted to the number of heavy trucks using the pavement every year and environmental factors: the water in the subgrade soil could freeze and heave and certain types of soils swell with the presence of water.

The previous paragraph provides us with a description of factors (elements) that contribute to explain the selected response. Now, we need to move through the steps of the estimation. The first step is to identify the elements that play a role, and this signifies the need to have one measurable indicator per element. So here is our list of elements: truck traffic, subgrade soil and pavement layer strength and weather (water). As it turns out, we could estimate the number of equivalent single axles (ESALs) of a given weight (80 kN or 18,000 lb) per year from traffic counts.

There is an expression called structural number developed by the American Association of State, Highway and Transport Officials that combines the thickness and quality of materials of the pavement into one that provides an indication of how strong the pavement is. Finally, there is an expression called Thornthwaite Moisture Index which reflects the weather of a given region by estimating its degree of water arriving to the soil. Table 5.1 shows the process of identification of factors and response we are trying to model.

The next step is to identify the nature of the relationship between the factors and the response. First, we need to think of the response moving in one particular direction, which is typically chosen as the undesired direction; in this case, an increase in deterioration is also an increase in IRI (Table 5.1).

Next, we identify the movement on each variable that produces an incremental effect on the response. For the first factor, an increase in the number of trucks will damage the pavement and so it will deteriorate more. So we could identify a direct response.

For us, a direct response signifies that an increase on the factor returns an increase on the response. For the second factor, an increase in pavement strength translates into a decrease in deterioration; the stronger the pavement, the less it deteriorates. We will call this type of relationship inverse. Finally, the more water reaching the pavement, the more it will deteriorate (from the effects of swelling and heaving), so the relationship in this case is direct as well.

To complete step 2, we simply associate *direct* relationships with a multiplicative effect and *inverse* effects with a division effect. So our relationship

TABLE 5.1

Factors That Contribute to Pavement Deterioration

Element	Description	Indicator	Relationship
Response	Deterioration	*IRI*	Incremental
Factor 1	Truck traffic	*ESALs*	Direct
Factor 2	Pavement strength	*SN*	Inverse
Factor 3	Presence of water	*m-index*	Direct

so far could take on a primitive functional form shown by the following equation:

$$IRI = \frac{(m - index)(ESALs)}{SN}$$

At this stage, we have completed step 2; however, we are presuming that all (direct and inverse) relationships are linear, which is somewhat a very strong assumption. So in order to relax this assumption, we introduce coefficients to raise each element. The calibration that proceeds on step 3 will confirm (coefficient = 1) whether or not the relationships were linear. Our functional form takes now the following form:

$$IRI = \frac{(m - index)^{\alpha}(ESALs)^{\beta}}{SN^{\delta}}$$

Notice that the response remains intact, and we do not apply any coefficient to it. Now, we concentrate on step 3. We need to calibrate the coefficients α, β and δ to replicate the response. Table 5.2 contains a sample of observations that we will use for this purpose.

TABLE 5.2

Sample Data Deterioration

IRI	*ESALs*	*SNC*	*m*
0.18	60,000	3.96	0.04
0.45	122,000	3.96	0.06
0.67	186,000	3.96	0.08
0.54	251,000	3.96	0.04
0.95	317,000	3.96	0.06
1.08	384,000	3.96	0.08
0.83	454,000	3.96	0.04
1.46	524,000	3.96	0.06
1.48	596,000	3.96	0.08
1.17	669,000	3.96	0.04
1.89	743,000	3.96	0.06
1.96	819,000	3.96	0.08
0.88	897,000	3.96	0.02
1.82	976,000	3.96	0.04
1.67	1,057,000	3.96	0.05
2.66	1,139,000	3.96	0.07
1.30	1,223,000	3.96	0.02
1.43	1,309,000	3.96	0.03
2.40	1,396,000	3.96	0.05
2.39	1,486,000	3.96	0.04

First, notice that we have developed one equation only. You have three coefficients (unknowns) and only one response, with one observation of the response per set of observed levels of the factors. So, we will proceed to run an ordinary least-square optimization to obtain the values of α, β and δ that minimize square differences between predicted and observed levels of IRI.

We can do this in Excel. First, we need to transfer the data from Table 5.2 into an Excel spreadsheet. Then we use the functional form created on step 2. Three additional cells in Excel will be used to contain the values of the coefficients α, β and δ. Then an additional column called predicted is added into Excel, and its value is coded to match the functional form along with the value of the unknown coefficients as shown in Figure 5.1.

Finally, the square difference between the observed value of IRI and the predicted value (using the functional form) is added into the spreadsheet. The total summation of the square differences becomes the objective of our optimization. There are however no constraints. As you can see, the problem is an unconstrained optimization, and the coefficients could take on any value.

SUM | × ✓ *fx* | =+(E2^K2)*(C2^K3)/(D2^K4)

	C	D	E	F	G	H	I	J	K
1	**ESALs**	**SNC**	***m***	**observed**	**predicted**	**diference^2**			
2	60000	3.96	0.04	**0.18**	**=+(E2^K**	**0.00**		alpha	0.609
3	122424	3.96	0.06	**0.45**	**0.40**	**0.00**		beta	0.781
4	186096	3.96	0.08	**0.67**	**0.67**	**0.00**		delta	6.060
5	251042	3.96	0.04	**0.54**	**0.55**	**0.00**			
6	317287	3.96	0.06	**0.95**	**0.85**	**0.01**			
7	384857	3.96	0.08	**1.08**	**1.17**	**0.01**			
8	453778	3.96	0.04	**0.83**	**0.88**	**0.00**			
9	524078	3.96	0.06	**1.46**	**1.25**	**0.04**			
10	595783	3.96	0.08	**1.48**	**1.65**	**0.03**			
11	668923	3.96	0.04	**1.17**	**1.19**	**0.00**			
12	743525	3.96	0.06	**1.89**	**1.65**	**0.06**			
13	819620	3.96	0.08	**1.96**	**2.12**	**0.03**			
14	897236	3.96	0.02	**0.88**	**0.98**	**0.01**			
15	976405	3.96	0.04	**1.82**	**1.59**	**0.05**			
16	1057157	3.96	0.05	**1.67**	**1.94**	**0.07**			
17	1139524	3.96	0.07	**2.66**	**2.53**	**0.02**			
18	1223539	3.96	0.02	**1.30**	**1.25**	**0.00**			
19	1309234	3.96	0.03	**1.43**	**1.68**	**0.07**			
20	1396642	3.96	0.05	**2.40**	**2.41**	**0.00**			
21	1485799	3.96	0.04	**2.39**	**2.21**	**0.03**			
22					**Total summation**	**0.43**			

FIGURE 5.1
Excel set-up.

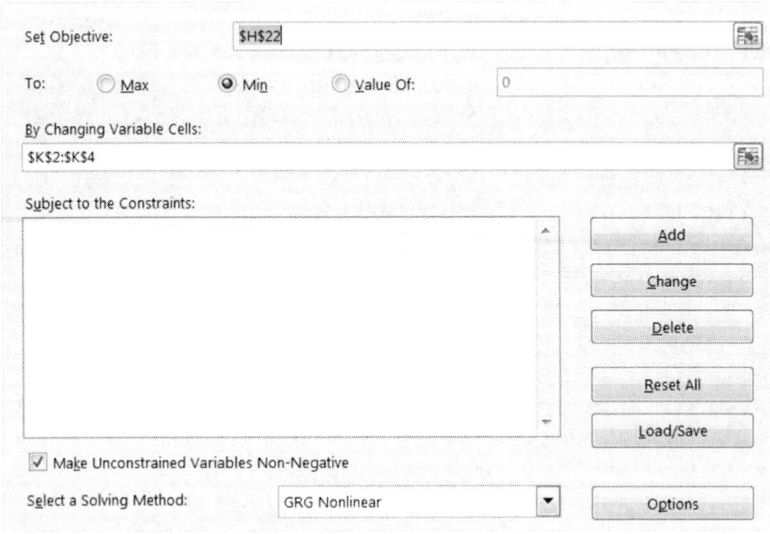

FIGURE 5.2
Excel's add-in solver set-up.

The reader must notice that before solving I set up the values of the coefficients to zero (you could have used the value of one); however, using zero allows the solver to depart from the non-contribution perspective, that is, a coefficient with a value of zero will have no contribution on the observed response.

Figure 5.1 shows the Excel set-up and the calibrated values from the optimization. Figure 5.2 shows the actual set-up on Excel's add-in solver used to find the calibrated values.

If you ever move into graduate school, then you will possibly learn how to develop systems of equations that are interrelated in which you could follow an iterative strategy to calibrate the coefficients (I do, however, present a couple of examples in Chapter 6). Deep understanding through years of research is typically devoted to develop such systems. Often, the equations you learn throughout your undergrad degree are the product of years of research, and as you will soon realize, the values of any constants on the equations are nothing more than calibrated values.

5.2.2 Example of Road Safety

We look into an example of road safety in which we want to build a functional form to estimate the contribution of several factors such as the number of lanes, density of intersections and others on collisions. Availability of data varies from government to government.

The three previous steps are repeated. First, we identify that there are many factors all related to collisions, traffic volume in the form of annual average daily traffic (AADT), number of lanes (x_1), density of intersections (x_2), width of shoulder (x_3) and radius of curvature(x_4).

We will not worry about step 1, that is, we will let the coefficients of our regression give us the indication of whether a factor has a direct (positive) or an inverse (negative) contribution to explain the response. For step 2, we will borrow the classical functional form shown in the following equation:

$$y = (AADT^{\alpha})e^{b_1x_1+\cdots+b_nx_n}$$

To conduct a linear regression, we need to warrant having a linear form, so we will apply natural logarithm on both sides of this expression, obtaining

$$\ln y = \alpha(\ln AADT) + b_1x_1 + \cdots + b_nx_n$$

If you simply think of $\ln AADT$ as a variable (say x_0) and the response $\ln y$ as a typical response (call it r), then the entire expression collapses into a linear form:

$$r = \alpha(x_0) + b_1x_1 + \cdots + b_nx_n$$

Given the fact that we have a linear expression, we can use a linear regression to estimate the value of the coefficients. For this, we turn our attention to STATA®. As learned before (Chapter 4), we can run a linear regression by writing the following in the command line: *regress* $rx_1x_2x_3\ldots$. Table 5.3 shows the observations used in this example.

The next step is to transfer the data from Excel (or any other database you maybe using) to STATA. The results from STATA are shown in Figure 5.3. As you can see, the number of lanes does not play a role on the explanation of collisions. The higher the traffic volume (AADT) and the higher the density of intersections, the more collisions we observe (both coefficients are positive), and the wider the shoulder, the less collisions we observe. The reader needs to remember that at a confidence of 95%, the corresponding p-values must be smaller or equal to 0.05. In this case, the p-value for x_1 goes way beyond such point and deems the number of lanes insignificant.

5.2.3 Estimation of Explanatory Power Standardized Beta Coefficients

This section presents the case of estimation when the explanatory power of the factors is needed for a specific purpose. If we use a simple regression to obtain the values of each beta, we will obtain one interpretation for traffic volume and other interpretations for the rest of the causal factors. Whenever you need to compare the contribution of elements to explain a response, you will need to standardize all the values on your database. This is achieved by taking each observation for a given factor, subtracting from it the mean of

TABLE 5.3

Sample Data Road Safety

y	x_0	x_1	x_2	x_3
28	8.78	2	0.5	0
6	8.77	2	0.17	2.67
22	8.78	2	0.5	0
133	9.23	1	0.63	1.13
162	9.13	2	0.8	0.47
2	8.28	2	0	1.80
5	8.28	2	0.17	0.45
23	9.17	2	0.17	2.5
119	9.11	3	0.7	0
162	9.08	3	0.85	0
46	8.27	3	0.43	1.89
3	8.03	2	0	1.5
13	8	2	0	1
8	8.2	2	0	0.45
5	8.2	2	0.33	1
20	8.14	2	0.5	1.4
17	8.3	2	0.5	2.33
2	8.46	2	0.17	2.23
3	8.47	2	0	1.7
3	8	2	0	2
11	8.05	2	0	1.5
3	8.05	2	0	2
12	8.04	2	0.71	0.86
7	6.9	2	0.17	0.65
9	6.9	2	0	0.65
3	7.97	2	0	0
2	7.97	2	0	1.05
2	7.97	2	0	0.53
3	7.78	2	0.33	1.42
4	7.33	2	0	0
2	7.13	2	0.17	1.25
6	7.13	2	0.33	0.93
3	7.14	2	0.33	1.25
2	6.92	2	0	1.2
6	6.23	2	0	0.6
6	6.77	2	0	1.03

```
. regress y x0 x1 x2 x3

      Source |       SS       df       MS              Number of obs =      36
-------------+------------------------------           F(  4,    31) =   16.03
       Model |  46635.1831     4  11658.7958           Prob > F      =  0.0000
    Residual |  22541.7892    31  727.154489           R-squared     =  0.6741
-------------+------------------------------           Adj R-squared =  0.6321
       Total |  69176.9722    35  1976.48492           Root MSE      =  26.966

------------------------------------------------------------------------------
           y |      Coef.   Std. Err.      t    P>|t|     [95% Conf. Interval]
-------------+----------------------------------------------------------------
          x0 |   18.63303   7.354041     2.53   0.017     3.634367     33.6317
          x1 |   6.237832   14.37443     0.43   0.667    -23.07901    35.55468
          x2 |    84.8366   21.58366     3.93   0.000     40.81643    128.8568
          x3 |  -16.71975   6.544035    -2.55   0.016     -30.0664   -3.373106
       _cons |  -139.3861   61.64367    -2.26   0.031    -265.1092   -13.66301
------------------------------------------------------------------------------
```

FIGURE 5.3
STATA® linear regression.

the factor and dividing by the standard deviation of the given factor under consideration.

Once you have done this, you can run a regression and obtain the standardized beta coefficients for each factor. This coefficients will read as the amount of variation that you need on each factor alone to produce a variation of 1 standard deviation on the response.

There is a built-in option on STATA to obtain the standardized beta coefficients automatically. In the command line after, you simply add the command *beta* after you specify the traditional regression: $regressyx_1x_2x_3, beta$.

The obtained values allow you to compare the explanatory power of each factor to explain the response. You can divide two beta coefficients and directly estimate how many more times one factor is capable of explaining the response. At this stage, you are probably wondering why and how you would ever use that. Here, I present an example that has been further addressed elsewhere by myself.

For instance, think of the warrant system for roadway lighting. The recommended coefficients used to make decisions for the provision (or not) of lighting are not calibrated, and for this reason they do not always work in the proper way when you sue this system. To make them work, you need to change them in such a way that their values reflect reductions in road collisions. This approach is called evidence based, and the aim is to have values that would be learned from crash history, that is, we will calibrate the scores of the warrant to values that warrant lighting when it will counteract night-time collisions.

For the calibration, we need to look into the explanatory power of each road deficiency to explain night-time collisions. That is, we have a series of geometric, operational and built environment elements that we can measure, and certain values of this elements represent higher degrees of deficiency of the road and then contribute to night-time collisions.

For instance, think of the width of the shoulder or width of the lane. If the shoulder is less than 2.5 m or the lane is less than 3.6, we step into deficient values. A lane of 2.5 m is so much more deficient than a lane of 3 m, so they could be blamed for producing collisions in dark circumstances at night when visibility is reduced and the driver expects a standard width lane (or shoulder that fully protects a stop vehicle). In a similar manner, we think of lighting as a countermeasure to help reduce the frequency and severity of collisions at night-time, counteracting the deficiencies found in other elements, that is, improving visibility.

To calibrate, we need to take the explanatory power of each element and use it to explain observed collisions. I present a very small database (in real cases, it contains hundreds, if not thousands, of segments) to illustrate this problem. A final word is important: when you obtain the standardized coefficients, you simply need to normalize them and rescale them in order to obtain the new values of the lighting warrant grid scores.

Table 5.3 (in the previous section) contains the information we will use for this example. We direct our attention to the STATA output once we include the *beta* command. Figure 5.4 illustrates the STATA output for the same example.

The contribution of x_0 (lnAADT) to explain the response is quite close (1.15 times) to that of x_3 (shoulder width), for the density of intersections (x_2) is about 1.8 times that of the width of the shoulder. With these values, it is easier to add them and obtain weights for their contribution, which will then be translated into scores. Table 5.4 shows the calculation of the scores. It is important to mention that the weights are multiplied by a factor of 20 in order to obtain scores. This number comes from the fact that the total combined contribution of the factors must add up to 20 and then is multiplied by another amount called classification points that we will not change ourselves.

Figure 5.5 shows the finally calibrated grid which only uses those significant elements in order to make decisions for the provision of roadway lighting. The reader can search for the original grid score for highways

```
. regress y x0 x1 x2 x3 ,beta

      Source |       SS       df       MS              Number of obs =      36
-------------+------------------------------           F(  4,    31) =   16.03
       Model |  46635.1831     4  11658.7958           Prob > F      =  0.0000
    Residual |  22541.7892    31  727.154489           R-squared     =  0.6741
-------------+------------------------------           Adj R-squared =  0.6321
       Total |  69176.9722    35  1976.48492           Root MSE      =  26.966

------------------------------------------------------------------------------
           y |      Coef.   Std. Err.      t    P>|t|                     Beta
-------------+----------------------------------------------------------------
          x0 |   18.63303   7.354041     2.53   0.017                 .3270272
          x1 |   6.237832   14.37443     0.43   0.667                 .0467698
          x2 |    84.8366   21.58366     3.93   0.000                 .5160159
          x3 |  -16.71975   6.544035    -2.55   0.016                -.2841406
       _cons |  -139.3861   61.64367    -2.26   0.031                        .
------------------------------------------------------------------------------
```

FIGURE 5.4
Standardized betas.

TABLE 5.4

Calibration of Scores

Element	Standardized Beta	Weight	Score
lnAADT (x_0)	0.327	0.290	5.80
Density intersection (x_2)	0.516	0.458	9.16
Shoulder width (x_3)	−0.284	0.252	5.04

Evaluation Grid (G1)									
Evaluated Element									
Length of Segment				Level (1, 2 or 3)					
Description of Analyzed Criteria		Real Value	Classification Points (PT)					Score (PD)	Scored Value = PD*PT
			1	2	3	4	5		
Geometry									
1	Total number of lanes		≤4	5	6	7	≥8		
2	Lanes width		>3.6	3.4 to 3.6	3.2 to 3.4	3.0 to 3.2	<3.0		
3	Median Width		>12	7.5 to 12	3.5 to 7.5	1.2 to 3.5	<1.2		
4	shoulder width		>3.0	2.5 to 3.0	1.8 to 2.5	1.2 to 1.8	<1.2	5.04	
5	Slope (from 0 to 7)		>6:1	6:1	4:1	3:1	<3:1		
6	Horizontal curve radius		>3500	1750 to 3500	875 to 1750	575 to 875	<575		
7	Vertical gradient		<3.0	3.0 to 4.0	4.0 to 5.0	5.0 to 7.0	>7.0		
8	Frequent interchange distance		>6.5	5.0 to 6.5	3.5 to 5.0	1.5 to 3.5	<1.5	9.16	
Subtotal									
Operational									
9	lnAADT → LOS		A	B	C	D	≥E	5.8	
Subtotal									
Notes: 1.Provision of lighting 2. Current speed: 80kph (95% of night-time operational speed if available, otherwise use the posted speed) 3. Development is defined based on the presence of commercial, industrial, or residential buildings. 4. Use the most deficient geometrical characteristics for road segments.								Grand Total	
								Required Scoring to provide lighting	60

FIGURE 5.5
Calibrated scores for roadway lighting.

(called G1) and realize that what we have done is simply the estimation of the explanatory contribution of each factor to explain collisions. We have used a standardized form in order to be able to express all contributions in common units. After this, the only remaining task is to modify the grid to contain only the significant factors, and the original values are changed to those estimated from the contribution to explain collisions, and hence, we end up with an evidence-based approach.

Some warnings are important to the reader. First, the reader must remember that a correlation analysis must be performed before selecting the causal factors that will be used for the analysis. Correlation typically takes the form of a matrix that identifies colinearities between factors and the response or simply between factors themselves, when the value of colinearity is high that signifies that the value should be dropped from the analysis.

The other element the reader must bear in mind is that more data and more elements should be used; however, sometimes data are not available and yet decisions need to be made. In this case, we can follow the same method proposed here with those elements available that survived the colinearity analysis and the significance test (through the p-values).

5.3 Predictions

The purpose of predictions is central to many models. Making predictions requires an ability to understand a system, which typically is captured by a series of mathematical algorithms that are able to represent reality (Figure 5.6).

Managers, planners and consulting engineers working for state/government clients will require the ability to forecast the future and to measure the impact of decisions (strategies, plans, policies, actions) on a given system of interest. Think, for instance, of the construction of a new metro

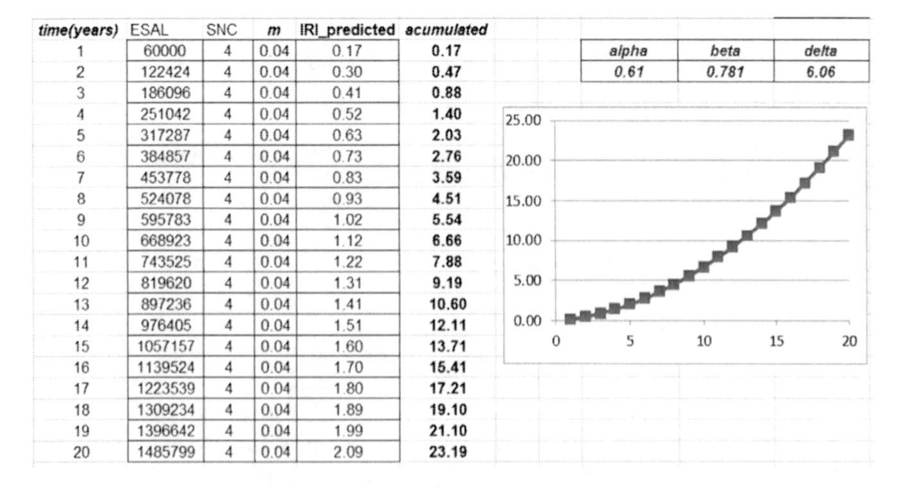

time(years)	ESAL	SNC	m	IRI_predicted	acumulated
1	60000	4	0.04	0.17	0.17
2	122424	4	0.04	0.30	0.47
3	186096	4	0.04	0.41	0.88
4	251042	4	0.04	0.52	1.40
5	317287	4	0.04	0.63	2.03
6	384857	4	0.04	0.73	2.76
7	453778	4	0.04	0.83	3.59
8	524078	4	0.04	0.93	4.51
9	595783	4	0.04	1.02	5.54
10	668923	4	0.04	1.12	6.66
11	743525	4	0.04	1.22	7.88
12	819620	4	0.04	1.31	9.19
13	897236	4	0.04	1.41	10.60
14	976405	4	0.04	1.51	12.11
15	1057157	4	0.04	1.60	13.71
16	1139524	4	0.04	1.70	15.41
17	1223539	4	0.04	1.80	17.21
18	1309234	4	0.04	1.89	19.10
19	1396642	4	0.04	1.99	21.10
20	1485799	4	0.04	2.09	23.19

alpha	beta	delta
0.61	0.781	6.06

FIGURE 5.6
Making predictions.

line; planners will be interested in finding out the impact of such construction to other modes of transportation. For this, a transportation engineer will identify those intersections affected by the construction of the metro (not all metro lines are underground, and even if they are underground, you need to build stations that are connected to the surface). The transportation engineer will be trying to predict the level of congestion during the closure of several roads as a consequence of the construction of metro stations (or metro lines). It is the job of this engineer to estimate current levels of congestion and to forecast the impact of the construction. Then the engineer turns his attention to mitigatory measures, temporary detours, temporary change of direction of some lanes or some roads, change in intersection control devices cycles, etc. For all this, the engineer needs the ability to measure current state, predict future condition and test the impact of several strategies.

The same is true for a municipal engineer trying to estimate the impact of water shortage. For this, the engineer needs to estimate current consumption, population growth and future demand and to test the impact of his decisions and several possible strategies. We will not however pursue this example here.

To predict, we need the mathematical representation created during the estimation. We require that such system has been previously calibrated. Finally, we need to extend such system to account for the time dimension. This part is tricky and depends on the type of system we have.

In some cases, the connection of our system can be achieved through a transfer function that takes current values and moves them into the future. In other occasions, the incorporation of the time dimension can be achieved directly into the very own equation(s) created during the estimation.

Let's look at the case of a very specific transfer function. Consider the condition of a bridge. Let's suppose you measure that in a Bridge Condition Index (BCI) that goes from 0 (very poor condition) to 100 (brand new). As the bridge deteriorates, the BCI drops. So every year the bridge condition reduces by a certain amount (call it D_t), but if you apply a maintenance, you could improve it (I). The amount of improvement will be considered fix for now. The transfer function between two periods of time is given by an expression that simply adds to the base period the amount of deterioration or the amount of improvement as follows:

$$BCI_t = BCI_{t-1} - D$$

$$BCI_t = BCI_{t+1} + I$$

Let's look now into a case in which the incorporation of the time dimension can be achieved directly into the equation. Consider again the condition of a pavement. If we introduce the base year level of condition IRI_0 and the time dimension to accumulate time into the environmental dimension and in the traffic loading dimension, we will achieve an expression that contains the time dimension and can be used for predictions:

$$IRI_t = IRI_0 + \frac{(mt)^{\alpha}(ESAL * t)^{\beta}}{SNC^{\delta}}$$

There are two concepts the reader needs to be familiar with: incremental and cumulative. The previous example of IRI is incremental; the estimation provides us with the amount of increase on IRI after 1 year. When we take such equation and incorporate the time dimension (as previously shown), we take the risk of ending up with an expression that is not properly calibrated. The way to solve this is by preserving the incremental nature of the system and simply adding every year values up until the time frontier to be explored. We will look into a way to predict for this example in the following text.

5.3.1 Example of Predictions

The amount of IRI that we are measuring so far is the one that we obtain after 1 year of deterioration, given the fact that ESALs are measured on a per year basis and the other factors don't contain the dimension of time. So what we are really measuring is the change in IRI from 1 year to another, but this amount should be added to the initial amount of IRI on a known base year in order to make predictions about future levels of IRI.

Let's now concentrate our attention to changing this problem into one in which we consider the time dimension. Let's take the case of a given segment of a road (instead of many segments as we did before), and let's attempt to evolve the functional form previously described into one that tells us the amount of deterioration for the next 20 years.

Let's presume we are looking into a segment with 60,000 ESALs per year, m-value of 0.04 and SNC of 5; as you know, these values won't change. The only value that will change is the total number of accumulated trucks if you accumulate them from year 1 to year 20.

There are other ways to make estimations and predictions; however, they escape the purpose of this book. I will content myself with mentioning the fact that the actual functional form is somewhat more complicated. You are supposed to test different functional forms until you find one that suits your observations and the phenomenon at hands. For the previous example of pavement deterioration, I could bring one classical form developed by Patterson and Attoh-Okine (1992). This form is presented in the following equation:

$$IRI_t = e^{mt}[IRI_0 + a(1 + SN)^{-5}NESAL_t]$$

As you can see, it takes on the accumulated number of ESALs from year 1 to year t, the structural number and the moisture index (m); however, it uses the base (initial) year roughness value (IRI_0), and it contains an extra coefficient a multiplying the accumulated number of ESALs. In addition, the moisture index is on the power of an exponent that contains the number of years elapsed since year 1. This expression seems more reasonable to represent the fact that every year the pavement undergoes a series of cycles of weather changes (frost–heave).

Exercises

1. Presume you have obtained the following coefficients: $\alpha = 265$, $m = 0.04$, $IRI_0 = 1.5$. Consider a structural number of 4.60425, ESALS on year 0 of 350,000 and growth of 3%. Estimate the amount of deterioration across time for 20 years. Use the following functional form:

$$IRI_t = e^{mt}[IRI_0 + a(1 + SN)^{-5}NESAL_t]$$

2. Estimate a coefficient to calibrate the following equation to the data in Figure 5.7.

ESAL	SN
25000	1.03
50503	1.18
76260	1.28
102275	1.354
128550	1.413
155088	1.463
181892	1.506
208963	1.545
236305	1.578

FIGURE 5.7
Exercise 5.2 making estimations.

$$\log(ESAL) = 2.717 + \log(SN + 1)^{\alpha}$$

ESAL is the total amount of traffic loads during the life span of the road. SN is the structural number; this value is provided on the following table. Estimate the value of the coefficient for the power of the term $SN + 1$.

3. Estimate standardized coefficients for the environment coefficient m, the $ESAL$ coefficient and the SNC coefficient shown in Figure 5.8. Interpret the standardized coefficients.
4. Use the following description to write a functional form. Add coefficients to each of the terms, and then use the data observed on the following table to estimate the value of the coefficients.

   Concrete strength s depends on the amount of cement c that you add to it. It also depends on the quality of the granular aggregates g.

   The other factor that we always consider is the amount of water w. If the amount of water goes beyond a certain ratio r, the strength of the mix drops. If the amount of water w drops below the minimum m, the strength of the mix drops as well. Do not use a coefficient for the water criteria. Presume $m = 0.1$ and $r = 0.3$. Figure 5.9 shows the observations that will be used for the calibration.
5. Estimate explanatory power of time, cost and comfort on utility for the data set in Table 5.5. Use STATA and the command *regress*.

IRI	ESAL	SNC	m
1.626044	25,000	3.964	0.04
1.72917	50,503	3.964	0.05
1.876007	76,260	3.964	0.06
2.076456	102,275	3.964	0.07
2.344782	128,550	3.964	0.08
2.001176	155,088	3.964	0.04
2.237347	181,892	3.964	0.05
2.551954	208,963	3.964	0.06
2.969642	236,305	3.964	0.07
3.525551	263,921	3.964	0.08

FIGURE 5.8
Exercise 5.3 making estimations.

MPA	c	g	w
15	0.2	5000	0.15
17	0.22	5000	0.15
20	0.25	5000	0.2
25	0.26	7500	0.2
32	0.28	7500	0.2
37	0.29	9000	0.2
40	0.33	12000	0.25
50	0.34	15000	0.25
60	0.35	20000	0.25

FIGURE 5.9
Data for Exercise 4.

TABLE 5.5
A Choice of Work Commute

Alternative	Time (Hours)	Cost ($/Month)	Comfort	Utility
1. Walk	4.5	10	3	5
2. Bike	1.18	10	4	6
3. Bus	1	101	1	5
4. Car	0.42	300	10	7
5. Walk–bus	0.82	101	0	6
6. Car–metro(1)	0.84	145	8	9
7. Car–metro(2)	1	305	9	7
8. Car–train(1)	0.92	175	7	6
9. Car–train(2)	1	175	7	5
10. Walk–train	1.25	125	6	4
11. Car–bus	0.67	151	5	5

Solutions

1. Figure 5.10 shows the detailed calculations associated to this specific problem.

   When we plot this information (Figure 5.11), we obtain a graph of IRI as the response and ESALs as the main driving factor. We can observe the speed of progression on deterioration measured through IRI.

	A	B	C	D	E	F	G
1	*time(years))*	IRI	ESAL	SNC	*m*	*alpha*	*IRI0*
2	1	1.64182546	350,000	4.60425	0.04	265	1.5
3	2	1.72888256	721,315	4.60425	0.04	265	1.5
4	3	1.82093704	1,103,769	4.60425	0.04	265	1.5
5	4	1.9182969	1,497,698	4.60425	0.04	265	1.5
6	5	2.02129021	1,903,443	4.60425	0.04	265	1.5
7	6	2.13026647	2,321,362	4.60425	0.04	265	1.5
8	7	2.24559805	2,751,818	4.60425	0.04	265	1.5
9	8	2.36768174	3,195,187	4.60425	0.04	265	1.5
10	9	2.49694045	3,651,858	4.60425	0.04	265	1.5
11	10	2.63382494	4,122,228	4.60425	0.04	265	1.5
12	11	2.77881575	4,606,710	4.60425	0.04	265	1.5
13	12	2.93242523	5,105,727	4.60425	0.04	265	1.5
14	13	3.09519973	5,619,713	4.60425	0.04	265	1.5
15	14	3.26772189	6,149,120	4.60425	0.04	265	1.5
16	15	3.45061317	6,694,408	4.60425	0.04	265	1.5
17	16	3.64453651	7,256,056	4.60425	0.04	265	1.5
18	17	3.85019915	7,834,552	4.60425	0.04	265	1.5
19	18	4.06835573	8,430,404	4.60425	0.04	265	1.5
20	19	4.29981153	9,044,131	4.60425	0.04	265	1.5

FIGURE 5.10
Solution for Exercise 5.1 making predictions.

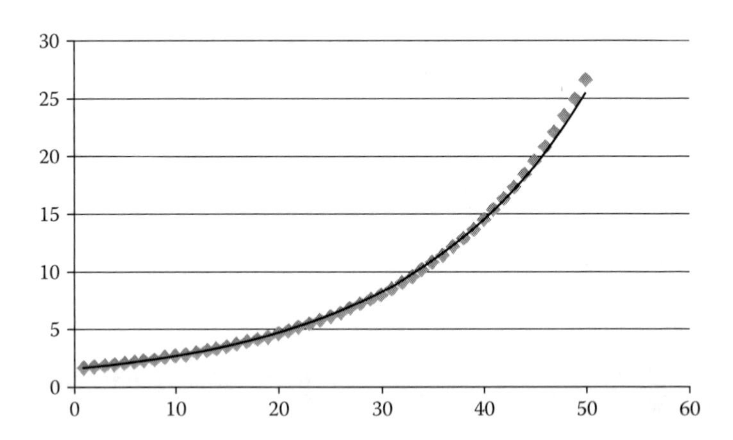

FIGURE 5.11
Solution for Exercise 5.1: IRI (m/Km) versus time (years).

2. We will assume that the bearing capacity of the soil is 30,000. If you notice the term in the logarithm containing the ΔPSI, it goes to $\log\left(\frac{1}{1}\right)$. You can also consolidate then all constants -7.67 and $2.32\log(M_R) = 2.32\log(30,000)$. The following equation shows a simplified version of the equation and will be used to estimate the value of the coefficient.

We will use the least-square approach to do the estimation. Figure 5.12 shows the observed and predicted values and then uses the least-square differences to estimate the power of the coefficient. Figure 5.13 illustrates the Excel interface used for the estimation.

As seen, the value of the coefficient is 6.2.

	A	B	C	D	E
1	ESAL	SN	log(ESAL)	Predicted	Difference
2	25000	1.03	4.398	4.624	0.051
3	50503	1.18	4.703	4.816	0.013
4	76260	1.28	4.882	4.937	0.003
5	102275	1.354	5.010	5.023	0.000
6	128550	1.413	5.109	5.089	0.000
7	155088	1.463	5.191	5.145	0.002
8	181892	1.506	5.260	5.191	0.005
9	208963	1.545	5.320	5.233	0.008
10	236305	1.578	5.373	5.268	0.011
11				difference	0.0929
12		alpha	6.201371089		

FIGURE 5.12
Solution for Exercise 5.2 estimation of coefficient predictions.

FIGURE 5.13
Solution for Exercise 5.2 Excel interface for the estimation.

```
. regress iri esal snc m ,beta
note: snc omitted because of collinearity

      Source |       SS       df       MS              Number of obs =      10
-------------+------------------------------           F(  2,      7) =   43.20
       Model |  2.88477881     2  1.4423894            Prob > F      =  0.0001
    Residual |  .233745859     7  .033392266           R-squared     =  0.9250
-------------+------------------------------           Adj R-squared =  0.9036
       Total |  3.11852467     9  .346502741           Root MSE      =  .18274

------------------------------------------------------------------------------
         iri |      Coef.   Std. Err.      t    P>|t|                     Beta
-------------+----------------------------------------------------------------
        esal |   5.51e-06   8.71e-07     6.32   0.000                 .7519697
         snc |  (omitted)
           m |   13.20837   4.694626     2.81   0.026                  .334495
       _cons |   .7144443   .2522885     2.83   0.025                        .
------------------------------------------------------------------------------
```

FIGURE 5.14
Solution for Exercise 5.3 estimating contribution power.

3. Figure 5.14 shows the results of the estimation using *STATA*. On the left side, we have the value of the coefficient and the p-value, and to the right we have the standardized coefficients. SNC is omitted because its value is fixed.

The coefficients from ESAL and m are both significant; the p-values are 0.000 and 0.025 correspondingly. The values of the coefficients are

MPA	c	g	w	predicted	min	r	obs-pred
15	0.2	5000	0.15	14.12003	0.1	0.3	0.774349
17	0.22	5000	0.15	14.55092	0.1	0.3	5.997976
20	0.25	5000	0.2	20.19943	0.1	0.3	0.039772
25	0.26	7500	0.2	29.99875	0.1	0.3	24.9875
32	0.28	7500	0.2	30.70817	0.1	0.3	1.668815
37	0.29	9000	0.2	36.8876	0.1	0.3	0.012634
40	0.33	12000	0.25	37.81788	0.1	0.3	4.761656
50	0.34	15000	0.25	47.1366	0.1	0.3	8.199044
60	0.35	20000	0.25	62.4289	0.1	0.3	5.899578
							52.34132
	alpha	0.315393					
	beta	0.944918					

FIGURE 5.15
Solution for Exercise 5.4 estimating α and β.

```
. regress utility time cost comfort

      Source |       SS       df       MS              Number of obs =      11
-------------+------------------------------           F(  3,      7) =    0.75
       Model |  4.61514779     3  1.5383826            Prob > F      =  0.5544
    Residual |  14.2939431     7  2.04199187           R-squared     =  0.2441
-------------+------------------------------           Adj R-squared = -0.0799
       Total |  18.9090909    10  1.89090909           Root MSE      =   1.429

------------------------------------------------------------------------------
     utility |      Coef.   Std. Err.      t    P>|t|     [95% Conf. Interval]
-------------+----------------------------------------------------------------
        time |  -.2500784   .5016773    -0.50   0.633    -1.436357       .9362
        cost |  -.0012313   .0083596    -0.15   0.887    -.0209986    .0185361
     comfort |   .2006834   .2187114     0.92   0.389    -.3164869    .7178536
       _cons |    5.30251   1.345463     3.94   0.006     2.120994    8.484026
------------------------------------------------------------------------------
```

FIGURE 5.16
Solution for Exercise 5.5 simple regression.

5.51E–6 and 13.20837. If you interpret these coefficients, the contribution power of the first one goes to zero. Once you estimate the beta coefficients, you can actually see that ESALs contribute to explain more (beta = 0.75) than the environment (with a coefficient of 0.33). This means that the ESALs are $\frac{0.7519697}{0.334495} = 2.248$ *times* more relevant than the environment to explain the deterioration of the pavement measured through IRI.

4. The fact that we have a minimum ratio r can be linked to the amount of water on a subtraction fashion such as $r - w$. If the amount goes beyond r, then the value of this term becomes negative. If the amount of water does not exceed such ratio, then the value is positive. However, there is also a minimum value m so having less than m is also counterproductive. We can represent this through another ratio $w - m$. This means that if the amount of water drops below the minimum, this amount becomes negative as well. We can then multiply both ratios $(w - m)(r - w)$. Figure 5.15 shows the solution.

   The equation obtained before calibration is the following:

$$s = c^{\alpha} g^{\beta} [(w - m)(r - w)]$$

5. The simple regression of time, cost and comfort on utility result in the following coefficients and p-values. Figure 5.16 shows the solution for the coefficients, although none of these was statistically significant.

   This suggests that there is no explanatory power on any of the three criteria (cost, time, comfort) to explain the level of utility.

References

1. Paterson, W. D. & Attoh-Okine, B., 1992. *Simplified Models of Paved Road Deterioration Based on HDM-III*. Proceedings of the annual meeting of the Transportation Research Board. Washington, DC.
2. StataCorp. 2011. *Stata Statistical Software: Release 12*. College Station, TX: StataCorp LP.

6

Land Use and Transport Models

This chapter provides the reader with an introduction to land use and transport models. The aim is to provide basic understanding of this type of models and the need for them to be integrated, rather than revising features of state-of-the-art models which are very complex. For this purpose, I concentrate my attention to two classical models used to generate and distribute trips: gravitational model and (multinomial) logistic model, and then I revise one method to predict the location of people for land use. This chapter also introduces the reader to the modeling features of an iconic model for integrated land use and transportation which is based on a matrix of origin–destination flows. Let us define first what is a land use and transport model and how it is important to engineers before we actually look into some of the modelling features.

6.1 Introduction

Land use models exist since the 1960s, and their purpose is to predict changes in population and build space development. Transportation models handle the task of estimating travel demand and its patterns over several modes of transportation (rail and subways, highways and roads, boats or planes). The idea behind land use and transport models is that travel originates as a product of people's demand, so changes in the location of people and development of new land (factories or residential areas) will impose new demand for freight or passengers on the network. This ideas are illustrated in Figure 6.1.

So the purpose of this type of models is to count with a linked or preferable integrated method capable of making predictions on land development, population growth, economic growth and travel demand for freight and passengers between regions or within metropolitan areas. We will cover soon two iconic models: the gravitational model, which is applicable to flows between regions, and a multinomial logistic approach, which is more naturally employed within a metropolitan region.

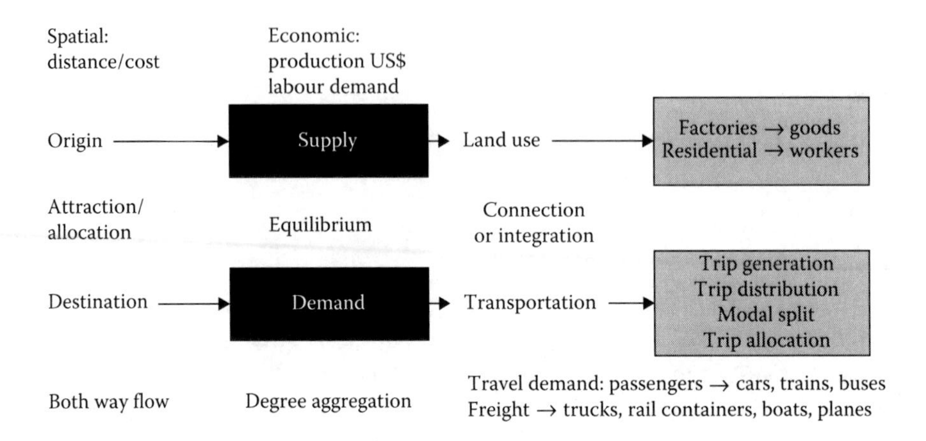

FIGURE 6.1
Elements of land use transport models.

A final word about these models is important. As any other model, they are abstractions from reality, which means that the model itself is a representation of reality based on one or various mechanisms that are believed to be able to capture the essence of the phenomena being modelled. For this purpose, the model needs to be calibrated to a known scenario, and once the model is capable of representing such scenario by reproducing some observed elements, then one is in capacity to use it to make predictions.

The usefulness to make predictions lays in the ability to forecast the future and test the implications of possible courses of action of decisions. It is common that such testing takes the form of what-if scenarios related to specific policies or planning decisions. For instance, a government maybe interested in knowing how many vehicles will enter the downtown area if they impose a congestion pricing scheme in which they charge vehicles entering the downtown core a premium fee during peak hours. The minister of transport maybe interested in knowing how many trucks would be using a new highway as opposed to a railway line connecting the industrial area with the main seaport and how this affects vehicle redistribution and land development around the new road. Governments can also test gas emissions and social implications.

You as an engineer, your entire career depends on clients that wish to design or build facilities, so the built environment and the transportation system will be always at the core of your career. At the beginning, perhaps you will be doing designs or studies, but eventually, you will take into decision-making roles as you advance in your future profession, so these tools will become part of your daily work, even if you hire a consultant to do the job; still, you need to understand the basics of the analysis, identify flows and be able to request changes that match your analysis or design needs.

6.2 Modelling: Estimation of Travel Demand with a Gravitational Approach

The first element that we will cover is the estimation of travel demand. The most iconic and basic model for this is an adaptation of the gravitational law of Newton, and hence it is called gravitational approach. It considers pairs of regions (or cities) and uses the principle of attraction between them to estimate the number of trips for an origin (i) and a destination (j).

$$F = \frac{Gm_1m_2}{d_{1,2}^2}$$

The gravitational approach predicts the size and direction of spatial flows (T), that is, of trips between origins i and destinations j. Each of them indexing many possible locations. Newton used the square of the distance ($d_{i,j}$); for us, the distance will be travel cost ($c_{i,j}$) raised to a power given by a parameter α to be estimated. Newton used a gravitational constant G; we will rather use a scaling constant (k) that will take care of converting nontrip units into travel demand. Such non-trip units are representative of the importance of a given region (W_i); you can think for simplicity of the size of the population. The following equation presents the general gravitational formula applied to predict trips between a pair origin–destination ($T_{i,j}$):

$$T_{i,j} = \frac{kW_iW_j}{c_{i,j}^{\alpha}}$$

So in the most general framework, the equation contains an indication of the attraction size for each pair origin i destination j, but very often we count with counts of trips exiting a given city i heading to all possible destinations j; this number can be incorporated into the gravitational equation. This allows us to start calibrating the model to what we have observed. First, notice how the summation over destinations gives us the total exit rate observed:

$$\sum_{j}^{n} T_{i,j} = O_i$$

Now consider a scaling constant given for origin i call it A_i. This scaling constant ensures that the summation of flows originated at i for all destinations j sum to the known amount O_i. Remember that W_j is the attractiveness of destination j (think of its population for simplicity):

$$A_i = \frac{O_i}{\sum_j^n O_i W_j c_{i,j}^{-}\alpha}$$

So the number of trips will be given by the scaling constant A_i, times the exit rate at the origin O_i, times the attraction pull of the destination W_j, divided by the cost raised to the power of a given constant α as seen in the following equation:

$$T_{i,j} = \frac{A_i O_i W_j}{c_{i,j}^{\alpha}}$$

The easiest way to see this in action is through a very simple example; presume you have only one origin (1) and three possible destinations (2, 3, 4). Figure 6.2 illustrates the situation and provides the solution.

I have used a simple Excel spreadsheet to solve this example. The inputs are distances between pairs which will be used as cost ($c_{i,j}$), population at destinations (W_j), number of units being produced and leaving the origin node 100 in this case and the constant $\alpha = 0.3$ to raise the power of the cost. The value of the scaling constant has been obtained on cell $B4$, and its explicit form is shown in Figure 6.3.

We may also have the number of trips arriving at a given destination j. Remember, you only need a traffic counter to record the number of trucks coming into town per year. This count can also be used to calibrate the model. Of course, we do not have the knowledge of where the trips originated, unless you have conducted an origin–destination survey, which is rare, and even if you did so, what you captured was on a given moment on time. Still you need

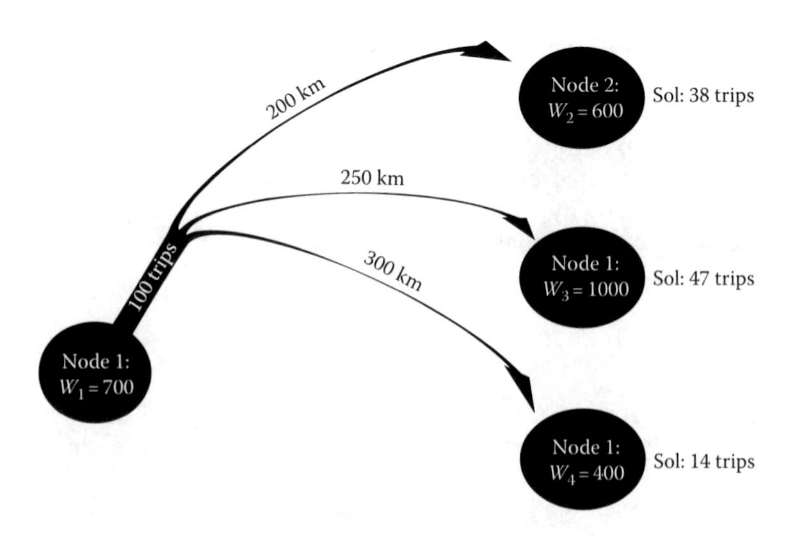

FIGURE 6.2
Production constrained gravitational example.

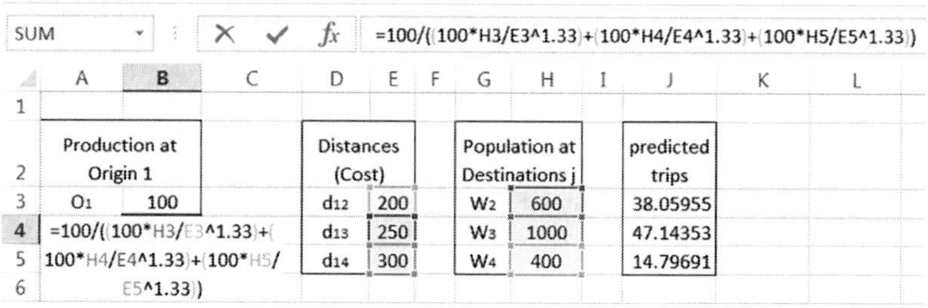

SUM | =100/((100*H3/E3^1.33)+(100*H4/E4^1.33)+(100*H5/E5^1.33))

	A	B	C	D	E	F	G	H	I	J	K	L
1												
2	Production at Origin 1			Distances (Cost)			Population at Destinations j			predicted trips		
3	O1	100		d12	200		W2	600		38.05955		
4	=100/((100*H3/E3^1.33)+(			d13	250		W3	1000		47.14353		
5	100*H4/E4^1.33)+(100*H5/			d14	300		W4	400		14.79691		
6	E5^1.33))											

FIGURE 6.3
Prediction of trips for production constrained.

the ability to encapsulate the phenomenon into a modelling framework. Call D_j the total number of trips arriving at location j:

$$\sum_{i}^{n} T_{i,j} = D_j$$

Now, consider a scaling constant given for destination j and call it B_j. This scaling constant ensures that the summation of flows originated at all i arriving at destination j sum to the known amount D_j. Remember that W_i is the attractiveness of origin i (think of its population for simplicity):

$$B_j = \frac{D_j}{\sum_i^n D_j W_i c_{i,j}^{-\alpha}}$$

So the number of trips will be given by the scaling constant B_j, times the arrival rate at the destination D_j, times the attraction pull of the origin W_i, divided by the cost raised to the power of a given constant α as seen in the following equation:

$$T_{i,j} = \frac{B_j D_j W_i}{c_{i,j}^{\alpha}}$$

Let's change the previous example and presume the old origin at 1 is now our destination and the three other locations are our new origins (2, 3, 4). The value of the scaling constant does not change; notice that the number of trips arriving at the destination is twice as much as those leaving the origin in the previous example, and even though this number plays a role on the scaling constant, its effect cancels out because you have its value on the numerator and denominator. Figure 6.4 illustrates this case and provides the solution;

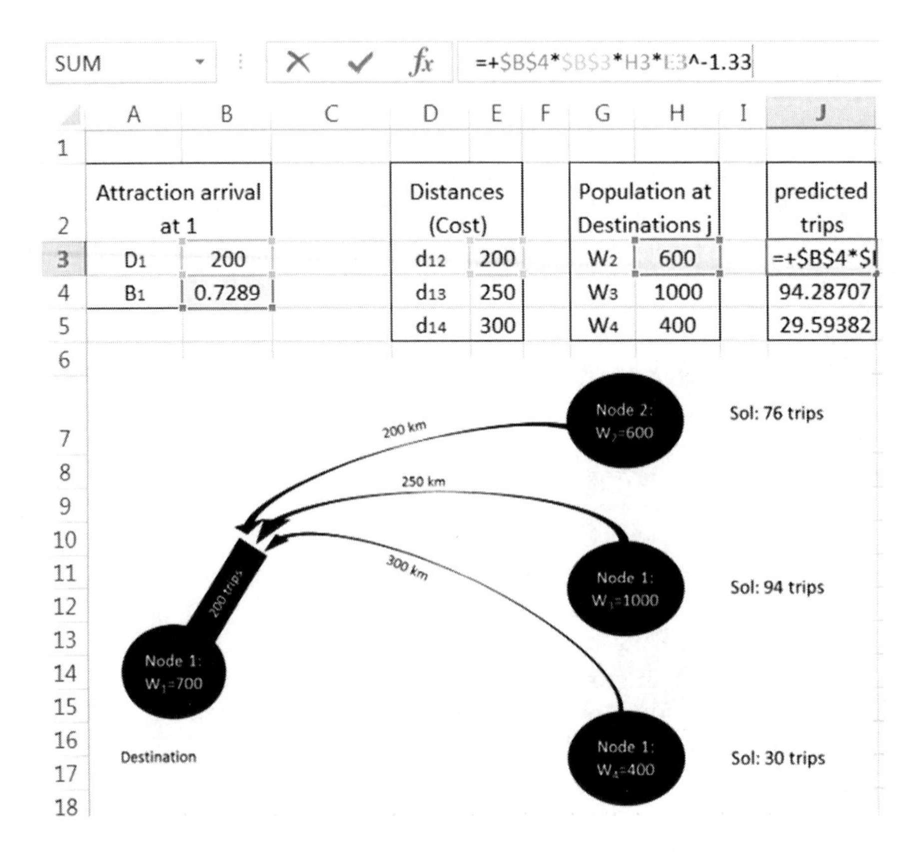

FIGURE 6.4
Prediction of trips for attraction constrained.

in this case, I have chosen to show the Excel codes behind the estimation of trips from 2 to 1.

Both constraints are commonly found together in what is called doubly constrained gravity model; hence, one must simultaneously satisfy

$$\sum_{j}^{n} T_{i,j} = O_i$$

and

$$\sum_{i}^{n} T_{i,j} = D_j$$

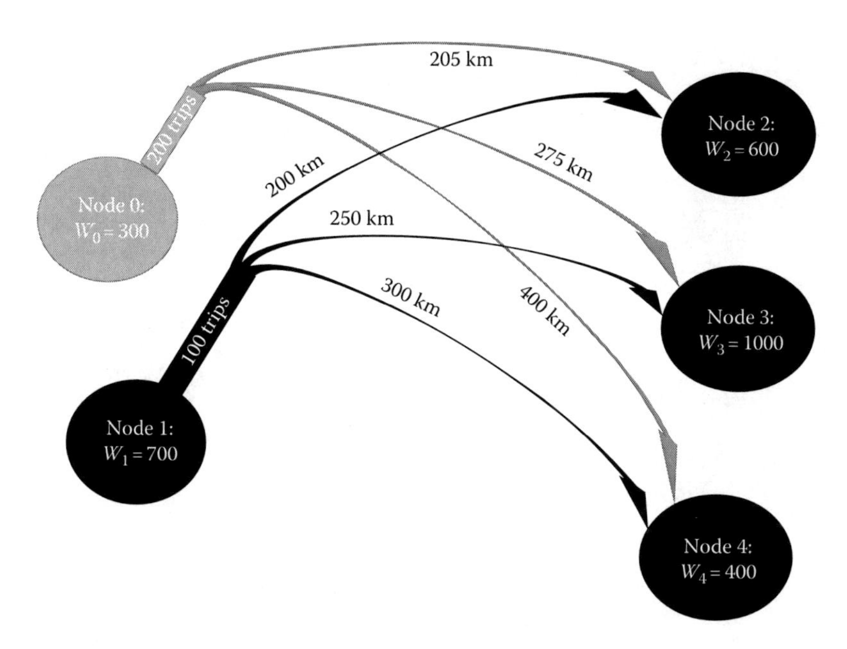

FIGURE 6.5
Doubly constrained.

Once this is incorporated in the gravity model, we obtain a doubly constrained estimation in which we have both scaling constants (whose calculation is as shown before):

$$T_{i,j} = A_i O_i B_j D_j c_{i,j}^{-}\alpha$$

Let us have an example with two origins (0 and 1) and three destinations (2,3,4). Figure 6.5 illustrates this case, and as you can see, it is the same example presented at the beginning of this chapter with an extra origin.

Populations ($W_{i,j}$), or any other measure of attraction, and distances – or any measure of transportation cost – are used to estimate the calibrating constants, shown in a simplified way later. One very important fact to notice is that the value of α will be iteratively iterated until the results of predicted flows match those observed by the following two equations:

$$A_i = \frac{1}{\sum_j^n W_j c_{i,j}^{-}\alpha}$$

and

$$B_j = \frac{1}{\sum_i^n W_i c_{i,j}^{-}\alpha}$$

	A	B	C	D	E	F	G	H	I	J	K	L	M
1	Production at Origin 0			Distances (Cost)			Population at Destinations j			predicted trips			
2	O_0	200		d_{02}	205		W_2	600		=+B3*B2*B14*B13/E2^L13			
3	A_0	3.7098		d_{03}	275		W_3	1000		99.7718273		T03	
4				d_{04}	400		W_4	400		21.0510519		T04	
5												**200.2**	
6	Production at Origin 1			Distances (Cost)			Population at Destinations j			predicted trips			
7	O_1	100		d_{12}	200		W_2	600		35.8280614		T12	
8	A_1	3.2192		d_{13}	250		W_3	1000		50.4302742		T13	
9				d_{14}	300		W_4	400		14.4809808		T14	
10												**100.7**	
12	Attraction arrival			Distances			Population at						
13	D_2	110		d_{02}	205		W_0	300		alpha		1.602	
14	B_2	4.9128		d_{12}	200		W_1	700					
15													
16	Attraction arrival			Distances			Population at						
17	D_3	150		d_{03}	275		W_0	300					
18	B_3	7.2502		d_{13}	250		W_1	700					
19													
20	Attraction arrival			Distances			Population at						
21	D_4	40		d_{04}	400		W_0	300					
22	B_4	10.455		d_{14}	300		W_1	700					

FIGURE 6.6
Solution to the doubly constrained gravitational problem.

The solution to this example after several iterations found that an $\alpha = 1.602$ was close enough to match exit flows of 200 leaving node 0 and of 100 leaving node 1. Figure 6.6 also illustrates the coding in Excel of the doubly constrained equation $T_{i,j} = A_i O_i B_j D_j c_{i,j}^{-}\alpha$.

6.3 Trip Distribution and Modal Split: Random Utility

The random utility approach is nothing else than a method used to learn the distribution of trips per mode and is linked to the observed characteristics of individuals and existing alternatives for transportation. In this sense, you will need to know the number of individuals at a given origin heading to a given destination. Think of those commuting from a residential area to downtown. Before I go any further, please notice that this presumes people had made their residence and work location choices already and this won't change in the future. In Section 6.4, we will learn how to estimate location decisions. For now, we complete our understanding of the modelling for travel before concentrating our attention to land use (location decisions and land development).

Individuals select those choices (transportation) that give them more satisfaction, that is, that have more value for them. The concept of value is typically referred in economic theory as utility. An individual selects alternative k if such alternative has more utility than any other one:

$$U_k > U_j \quad \forall j = 1, \ldots, n,$$

We really don't care about learning the actual values of utility, but rather we care about the choices made by individuals i, that is, the probability of that individual i selecting a given alternative k which itself comes from his perceived utility:

$$P_{i,k} = P(U_{i,k} > U_{i,j})$$

Utility can be broken down into a fix component and a random error term. The fix component can be estimated from observed characteristics of the choices and the individual. This is especially true since single individuals will give more value to use their car to commute than married individuals with children. So the choice is not only related to the alternative characteristics (travel time, cost, comfort, etc.) but also the individual characteristics (income, kids, etc.).

Let us call $V_{i,k}$ the observable term of the random utility of individual i selecting choice k. As you can see, what we are doing with the index is matching individuals to choices. For this, we use the probability again, but this time in terms of a fixed term of the utility, which is based upon the observable attributes, and a random term $E_{i,k}$ of utility, which we do not observe:

$$P_{i,k} = P(V_{i,k} + E_{i,k} > V_{i,j} + E_{i,j}).$$

The only thing left now is to express the decision choice as a function of choosing one alternative over all other alternatives. That is, we obtain a weight of each alternative among the others. Suppose we use the Weibull distribution. This will result in using the exponential to obtain the weights.

$$P_{i,k} = \frac{\exp(V_{i,k}(X_k, S_i))}{\sum_j^n \exp(V_{i,k}(X_k, S_i))}$$

This expression assumes that the individual is aware of all possible choices, for now lets go with this assumption. Given the fact that we do not observe the fix utility, it is rather simpler to let those factors that play a role in the individual's decision identify the probability. The factors are represented by a vector X_k of observed attributes for the alternative and a vector of attributes S_i of observed characteristics for the individual. We will use the statistical package *STATA*® to estimate the probabilities. *STATA* has a built-in command for logit estimation. The logit estimation simply employs the previous formula to estimate the probability for choice k. Let's look at an example.

6.3.1 Example: Logit

Suppose you have observed characteristics and choices for 20 individuals (this small size is only intended for academic purposes; typically, you need much bigger sample sizes).

Table 6.1 shows the agents (A) and their characteristics: income (I) in thousands per year, has kids (K), owns a car (O) as well as the choice attributes of travel by bus cost (B) or travel by car cost (C) to work and the travel time by bus (T). The choice of travelling by bus Y, having kids and owning a car is represented by a binary variable: yes = 1, no = 0.

To run *STATA*, we simply use the command *logit y i k o b c t*. Notice that the sequence of commands follows a given order; first, the type of model 'logit', then the response 'y' and then the attributes in any order (although I follow the same order as in the table). *STATA* automatically identifies the

TABLE 6.1

Bus or Car?

A	*I*	*K*	*O*	*B*	*C*	*T*	*Y*
1	20	1	0	150	300	60	1
2	100	1	1	150	300	40	0
3	45	1	1	150	300	45	1
4	75	1	1	150	300	45	0
5	50	0	1	150	200	50	0
6	60	0	1	150	200	60	0
7	35	0	0	150	300	60	1
8	80	1	1	150	300	80	0
9	40	1	1	150	200	50	1
10	55	1	1	150	300	45	1
11	70	1	0	150	200	30	1
12	65	0	1	120	300	40	1
13	25	0	1	120	350	55	1
14	85	0	1	120	325	40	0
15	90	0	0	120	400	35	1
16	52	0	1	120	200	45	0
17	37	0	1	150	200	35	1
18	67	0	1	150	250	35	0
19	56	1	1	150	200	55	1
20	88	1	1	150	200	45	0

```
. logit y i k o b c t

note: o != 1 predicts success perfectly
      o dropped and 4 obs not used

Iteration 0:   log likelihood = -10.965027
Iteration 1:   log likelihood = -2.3891148
Iteration 2:   log likelihood = -.97884697
Iteration 3:   log likelihood =          0
Iteration 4:   log likelihood =          0

Logistic regression                           Number of obs   =         16
                                              LR chi2(-1)     =      21.93
                                              Prob > chi2     =          .
Log likelihood =          0                   Pseudo R2       =     1.0000

------------------------------------------------------------------------------
           y |      Coef.   Std. Err.      z    P>|z|     [95% Conf. Interval]
-------------+----------------------------------------------------------------
           i |  -6.230316          .       .       .            .           .
           k |   59.98421          .       .       .            .           .
           o |  (omitted)
           b |   .6869601          .       .       .            .           .
           c |   1.559285          .       .       .            .           .
           t |   .7038294          .       .       .            .           .
       _cons |  -137.9599          .       .       .            .           .
------------------------------------------------------------------------------
Note: 9 failures and 7 successes completely determined.
```

FIGURE 6.7
Selecting bus to commute to work.

redundancy on using auto ownership as an attribute and removes it from the analysis.

STATA is capable of successfully predicting 16 out of 20 observations. The results from *STATA* show the following coefficients for each of the attributes: −6.23 (income), 59.98 (has kids), 0.69 (cost of bus ride), 1.55 (cost of car ride) and 0.70 (travel time). Figure 6.7 shows the results from *STATA*.

As you can see, having a kid is the major factor in explaining the use of the bus or not. Having a kid can influence the choice of bus for several reasons: It is preferable to leave the car home in case they need it, having a kid implies expenditures, and it also signifies being very busy when at home, so why stress out driving back when you can ride?

The previous result gives you the ability to forecast between two modes (bus or car). The logistic regression is binary (yes or no). However, it does not enable you to make choices between many competing modes. We will cover now an example of multinomial logit which allows you to do that.

6.3.2 Example: Multinomial Logit

Multinomial logit follows the same logic as the logistical regression or logit model, but it applies to many categories. Let's take the example of the previous table and modify the choices to be $bus = 1$, $car = 2$ and $bike = 3$. For simplicity, I have dropped some attributes and preserved only income I, kids K and travel time by bus T.

y	Coef.	Std. Err.	z	P>|z|	[95% Conf.	Interval]
1	(base outcome)					
2						
i	.0893949	.0514566	1.74	0.082	-.0114582	.1902479
k	-2.202523	1.586833	-1.39	0.165	-5.31266	.9076128
t	.1916946	.1442783	1.33	0.184	-.0910857	.4744749
_cons	-13.07815	8.520831	-1.53	0.125	-29.77868	3.622368
3						
i	-1.133994	273.653	-0.00	0.997	-537.484	535.216
k	-8.76757	7339.648	-0.00	0.999	-14394.21	14376.68
t	-1.596295	624.3169	-0.00	0.998	-1225.235	1222.042
_cons	114.5456	24034.65	0.00	0.996	-46992.51	47221.6

FIGURE 6.8
Multinomial logit, base = bus trip.

To run *STATA*, we simply use the command *mlogit y i k t, base(1)*. The sequence of commands follows: the model 'mlogit', the response 'y' and the attributes along with a definition of the base level for the response, meaning that the coefficients are all based on comparisons to such base level. *STATA* results (Figure 6.8) show that higher income (0.089) and travel time (0.19) will encourage travelling by car as opposed to bus, having kids will discourage it (−2.20). These results are significant at 80% confidence interval (C.I). For biking, we observe non-significant results; remember, we have a very small set of data.

The previous result gives you the ability to forecast how many trips will be made per mode of transportation (bus, car, train, etc.). The model will take the entire population along with its characteristics and predict how many individuals ride the bus, and then you would know that a bus can carry 60 individuals, so these two elements will give you the number of buses. Similarly for trains, we will know that trains carry five waggons with a capacity of 90 individuals each waggon. A car could be presumed to carry on average 1.5 individuals (carpooling). So at the end of the day, the gravitational approach tells me how many trips in total I observe between two regions, and the random utility allows me to distribute trips per mode of transportation. Now, we turn our attention to the estimation of where people live, which relates to the land use component of this chapter.

6.4 Modelling: Location and Land Development

The most iconic model for location is the Lowry (1964) model developed under the assumption that employment will condition the place of residence. Employment is divided in basic (manufacture) and non-basic (services). Basic employment is exogenously (external to the model) supplied and non-basic is derived within the framework. Both are used to estimate the location of residents. This residential location is used to estimate service employment; the model proceeds iteratively assigning activities to urban locations.

Basic employment is externally determined given the fact that such type of industries requires specialized places to settle and export their goods to external markets; hence, they do not play a role in the location of residents. On the contrary, services locate close to urban areas because their goods are demanded by residents of such areas.

The decisions to locate in a given area are related to service and industry employment. People will locate in an origin area based on their places of employment, considering two main factors: how attractive a zone is and the distance to his or her place of employment. Attractiveness of a given zone i (W_i) relates to availability of shops, leisure and recreation facilities, house/rent price levels and availability of clinics/schools. Distance relates to the cost ($c_{i,j}$) to commute (time, money, comfort, convenience).

Service companies will locate as close as possible to residential zones to provide good accessibility to people. The Lowry model goes through four major components in an iterative fashion.

Step 1. Basic (E_j^b) and service employment (E_j^s) are combined to estimate total employment in a given zone j. Service employment is typically set to zero for the first iteration.

$$E_j = E_j^b + E_j^s$$

Step 2. Total employment E_j is used to estimate the number of residents that lives in an origin zone i and work in a destination j called $R_{i,j}$. A parameter called B_j is used to make sure that the correct number is assigned to zone and i, that $R_{i,j} = E_j\mu$. Two coefficients are used to calibrate the model: β and α. The first one serves to scale the cost and regulate the order of magnitude on the predicted number of residents $R_{i,j}$; if cost is expressed in dollars and then β will take on a small value in the order of thousands of decimals. The other coefficient α will be used to adjust each $R_{i,j}$.

$$R_{i,j} = E_j \mu B_j W_i^{\alpha} \exp(-\beta^r c_{i,j})$$

The strategy I recommend is to set up the problem and use the value of β^r to adjust the predicted employment ratio to match the observed employment ratio of the given zone with respect to the others. Then, change α to modify the scale of prediction to match the observed scale of magnitude; this will affect the employment ratio that you observed, so you would have to fine-tune β^r again and then α. It is possible that after three iterations you would reach good values of α and β.

Step 3. Destination zones j are allocated service employment (E^s) based on places of residence i (origin zones from step 2).

$$E^s_{i,j} = R_i s A_i W_j^{\alpha} \exp(-\beta^s c_{i,j})$$

In this context, $R_{i,j}$ is the estimated number of residents of origin i who work in j. R_i is the total number of residents from zone i, s is a service employment-to-population ratio and A_i is a constant that ensures the correct allocation of service employees to zone j.

Step 4. Go back to step 1. Newly calculated service employment is added back to the basic employment (fixed–exogenous). At each iteration, the number of service employment is expected to reduce and eventually converges to zero.

6.4.1 Basic Example of Lowry's Model

Suppose you have one residential zone, one downtown area and a commercial zone. This is the case of some small towns in rural areas of Canada. Our total working-age population is 5000 people with basic employment of 1000. There are two working areas, downtown (i = 1) and industrial/commerce (i = 2), and two residential areas (i = 3 and i = 4) as shown in Figure 6.9. Service employment is set to zero.

The set-up for steps 1 and 2 is shown in Figure 6.10. We concentrate now in advancing this problem to step 3 in which destination zones j = 1 and j = 2

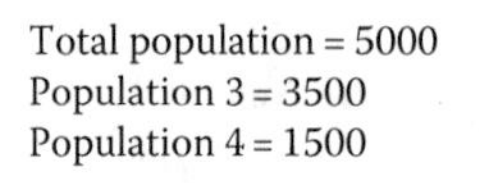

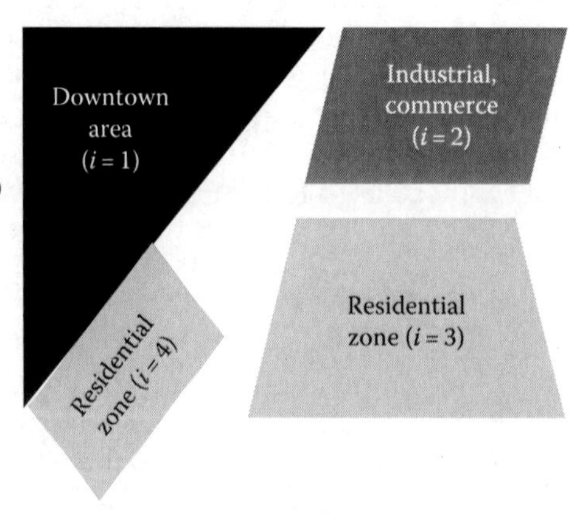

FIGURE 6.9
Our metropolitan area.

	A	B	C	D	E	F
1	- Basic Employment	E_1^b	1000		E_2^b	600
2	- Service Employment	E_1^s	0		E_2^s	0
3	Total Employment	E_1	1000		E_2	600
4	*Observed employment ratio*			1.66667		
5	beta regulate cost	β^r	0.54	residents		
6	Cost c_{ij}	c_{31}	20	$/day		
7		c_{32}	10	$/day		
8		c_{41}	5	$/day		emplo
9		c_{42}	13	$/day		
10	Attractiveness	W_3	7	index 0-10		
11		W_4	5	index 0-10		
12	calibration coefficients	alpha	2.43	regulates attactiveness		
13		β^s	0.5	regulate cost on services		
14		Ai	1			
15		μ	1.66	populat-employm.ratio		
17	STEP 2		Predict	Observed	Destination	
18	R_{ij}	R31	2.7	1005.6	j = 1	
19	Population from i	R41	1003.0			
20	working on j	R42	13.3	607.1	j = 2	
21		R32	593.75			
22	*Predicted Employment Ratio*			1.65649		

FIGURE 6.10
Calibrating α and β.

are allocated service employment (E^S) based on places of residence i (zones 3 and 4 from step 2). For this purpose, we use the following equation that is already explained before:

$$E^S_{i,j} = R_i s A_i W_j^{\alpha} \exp(-\beta^s c_{i,j})$$

Figure 6.11 shows the complete set up for the first iteration. As seen, values are aggregated per origin and destination. Some calibration coefficients are used to obtain employment ratios and to adjust the order of magnitude to the observed one.

I won't pursue iterations of this approach but the reader should know that the final estimation of location of people at residential zones will be achieved; this will tell us how many people we have at each region, and this will be an indication of land development in the form of subdivisions to satisfy any demand for housing product of the new employment created.

	A	B	C	D	E	F	G	H	I	J
1	- Basic Employment	E_1^b	1000		E_2^b	600		Population R_i		**B coeff**
2	- Service Employment	E_1^s	0		E_2^s	0		R3	3500	**0.7**
3	Total Employment	E_1	1000		E_2	600		R4	1500	**0.3**
4	*Observed employment ratio*			1.66667				Total	5000	
5										
6	Observed popul-empl.	μ	1.66	populat-employm.ratio						
7		c_{31}	20	$/day			STEP 3			
8	Cost c_{ij}	c_{32}	10	$/day			$\beta^{s=}$	0.150	regulate cost on services	
9		c_{41}	5	$/day			Ai=	0.054		
10		c_{42}	13	$/day			s=	0.600	% Service Worker / all workers	
11		W_3	7	index 0-10						
12	Attractiveness	W_4	5	index 0-10						
13	CALIBRATION COEFF	alpha	2.43	regulates attractiveness			R3	596.4	obtained from before	
14	CALIBRATION COEFF	β^r	0.54	regulate cost for residents			R4	1016.3		
15								0.586868		
17	**STEP 2**		Predict	Observed	Destination					
18	R_{ij}	R31	2.7	1005.6	j = 1		E32	487.8124	596.658	0.5902575
19	Population from i	R41	1003.0				E31	108.8456		
20	working on j	R42	13.3	607.1	j = 2		E41	776.8584	1010.844	
21		R32	593.75				E42	233.9852		
22	*Predicted Employment Ratio*			1.65649						

FIGURE 6.11
Set-up for steps 3 and 4.

6.5 Advanced Land Use and Transport Systems

The most advanced systems for land use and transportation take the same ideas behind the gravitational model but extend them to many sectors and goods across regions. This takes the form of detailed input–output matrices that could start with aggregated sectors (industry, commerce, land, employment, etc.) and could be disaggregated to any level of desired detail (bread producers, cereal producers, meat producers, etc.). This amount is used to constrain the model to known flows in order to calibrate the coefficients.

They also make use of multinomial logit in order to simulate decisions for modal split and trip distribution (using transport cost and other attributes). A multinomial logit is also used to simulate decisions for the location of activities (residences for people and floorspace for industries) which leads to land values and transport cost. Floorspace does not contain an element of transportation its estimation.

Let's take TRANUS (de la Barra 1992) to continue our explanation. As shown in Figure 6.12, it estimates increments on basic activities and floorspace by looking into previous period versus current period. Such increments trigger demand for floorspace and generate new transportation flows.

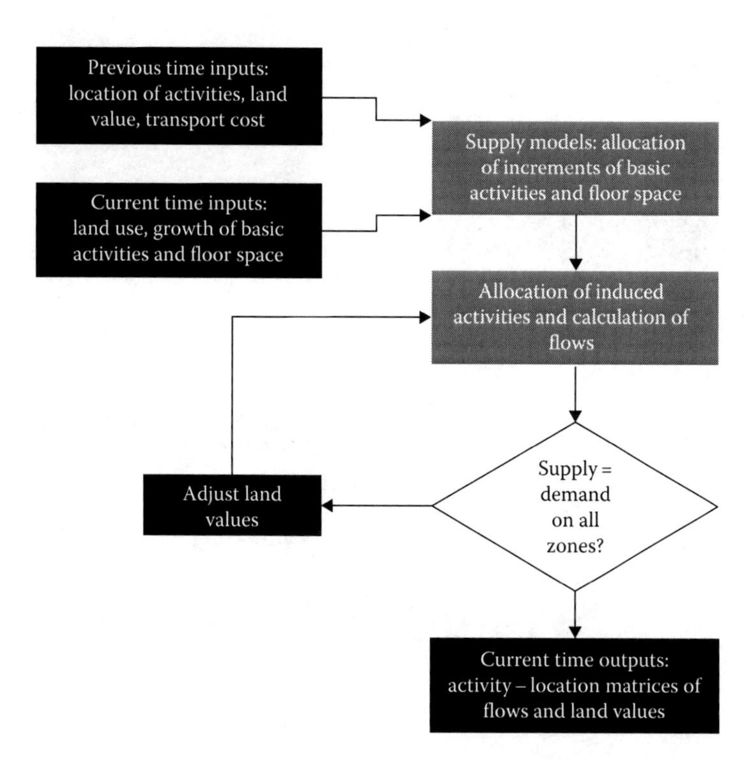

FIGURE 6.12
TRANUS inputs/outputs.

At the end, the system iterates until supply meets demand (equilibrium) in all zones. Then the system moves 1 year into the future.

In the sense of location decisions, the system expands upon Lowry concepts and allow for basic activities to be allocated space. From a pure modelling perspective, the model starts at the regional level with some economic information about employment and population, which is the basis of Lowry's model. Then it moves into two components: one for the location of basic employment and another one for the location of residents and service employment. Finally, it proceeds into a well-known four-step modelling using multinomial logit (Figure 6.13).

The feedback link between the activities (land use) and the transport (travel demand) comes in the form of transport cost and reduction of utility from congestion. The system is given a representation of capacity for each link and mode, and a measure of how sensitive the choice of a mode is based on an elasticity. Then if the demand reaches capacity, the waiting times for public transportation (or travel time for freight) will increase, making such route/link less attractive. This increases transport cost which is

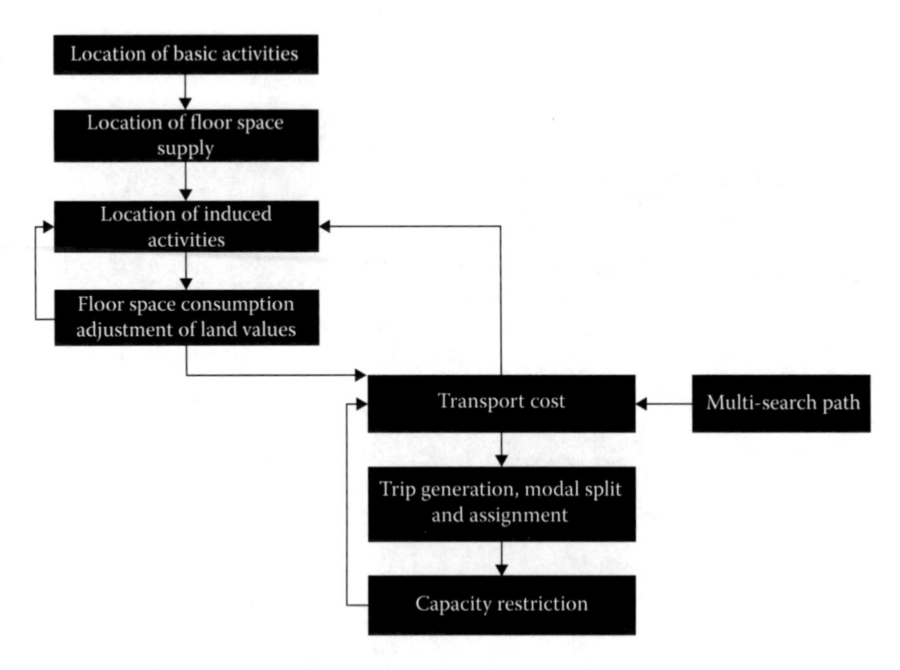

FIGURE 6.13
TRANUS mechanisms.

an element on the multinomial logit model used to obtain the modal split and trip distribution, hence producing a reallocation of trips to alternate links/modes.

Exercises

Given the nature of the problems at hand, I provide the solution next to each problem. This change from previous chapters is given to facilitate the reader's problem-solving skill development, because even with the provided solution, the reader will have to work out the development of a spreadsheet to be able to solve.

1. Formulate a gravitational model in Excel for the following destination with three origins and estimate the scaling constraint and the three flows originating at 2, 3 and 4 (Figure 6.14).
2. Formulate a gravitational model in Excel for a system with two origins (0 and 1) and three destinations (2, 3 and 4) and estimate the scaling

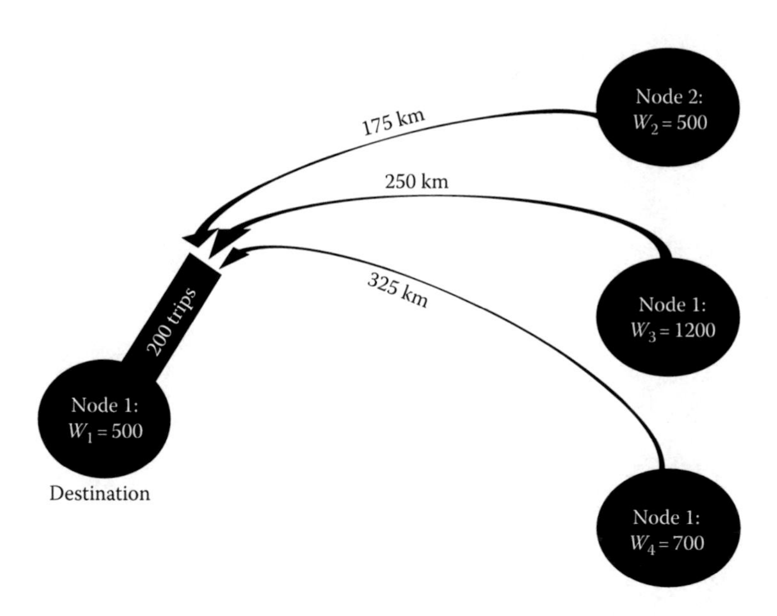

FIGURE 6.14
Exercise 6.1.

constants and the three flows originating at 0 and 1 going to 2, 3 and 4. The add arrival flows; suppose you just finished modelling and learned the observed arrival at 2, 3 and 4 are 120,142 and 38, correspondingly. Would you change your model?

3. Take the values given in Table 6.2 and replace all trips made by bike for trips made by bus, and then change the values made by bus with a value of 1 and by car with a value of 0. Run a logit binary model on any statistical software and interpret the results (Figures 6.5, 6.15 through 6.17).

   Remember that the coefficients are for bus travel as compared to car travel. Higher income (−0.96) and travel time (−0.208) negatively impact bus travel; having kids (2.27) has a strong contribution to select bus trips. All of them have an 83.5% confidence interval (based on the worst case for kids with p-value of 0.165).

4. Use a multinomial logit to estimate the explanatory power of cost, travel time and comfort to explain utility. Table 6.3 gives you the observed data.

Solutions

The solution for Exercise 6.4 is Figure 6.18, the level of utility 5 has been set as base. This means that for utility level 4, the time and cost exhibit negative

TABLE 6.2
Bus, Car or Bike?

A	*I*	*K*	*T*	*Y*
1	20	1	30	3
2	100	1	40	2
3	45	1	45	1
4	75	1	45	2
5	50	0	50	2
6	60	0	60	2
7	35	0	25	3
8	80	1	80	2
9	40	1	50	1
10	55	1	45	1
11	70	1	30	1
12	65	0	40	1
13	25	0	40	3
14	85	0	40	2
15	90	0	35	1
16	52	0	45	2
17	37	0	35	3
18	67	0	35	2
19	56	1	55	1
20	88	1	45	2

	A	B	C	D	E	F	G	H	I	J
1										
2	Attraction arrival at 1			Distances (Cost)			Population at Destinations j			predicted trips
3	D_1	200		d_{12}	175		W_2	500		64.34981
4	B_1	0.6191		d_{13}	250		W_3	1200		96.10343
5				d_{14}	325		W_4	700		39.54677

FIGURE 6.15
Solution to Exercise 6.1.

returns and comfort a positive contribution; a similar trend can be observed for utility levels 6 and 9, and the only exemption is on utility level 7. However, the *p*-values of all of this results are very poor, deeming the analysis and interpretation non-significant. The reader should bear in mind that this possibly comes from the fact that we are only using 11 observations. A larger database could solve this issue.

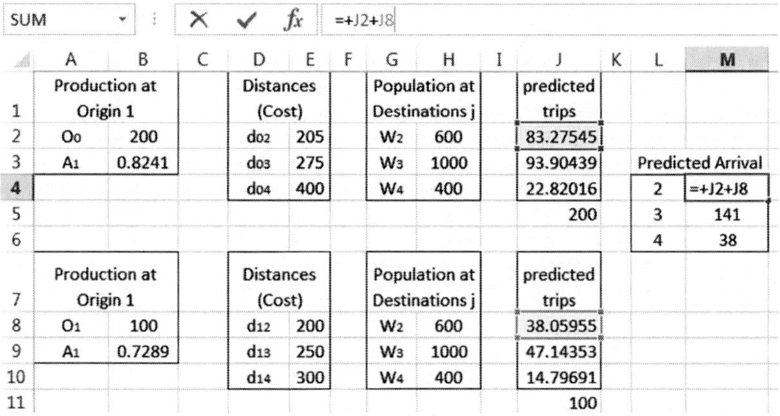

SUM | × ✓ *fx* | =+J2+J8

	A	B	C	D	E	F	G	H	I	J	K	L	M
1	Production at Origin 1			Distances (Cost)			Population at Destinations j			predicted trips			
2	O_0	200		d_{02}	205		W_2	600		83.27545			
3	A_1	0.8241		d_{03}	275		W_3	1000		93.90439		Predicted Arrival	
4				d_{04}	400		W_4	400		22.82016		2	=+J2+J8
5										200		3	141
6												4	38
7	Production at Origin 1			Distances (Cost)			Population at Destinations j			predicted trips			
8	O_1	100		d_{12}	200		W_2	600		38.05955			
9	A_1	0.7289		d_{13}	250		W_3	1000		47.14353			
10				d_{14}	300		W_4	400		14.79691			
11										100			

FIGURE 6.16
Solution Exercise 6.2.

```
Logistic regression                             Number of obs   =         20
                                                LR chi2(3)      =      13.19
                                                Prob > chi2     =     0.0043
Log likelihood =  -7.169661                     Pseudo R2       =     0.4791

------------------------------------------------------------------------------
           y |      Coef.   Std. Err.      z    P>|z|     [95% Conf. Interval]
-------------+----------------------------------------------------------------
           i |  -.0967163   .0476824    -2.03   0.043    -.1901722   -.0032604
           k |    2.27364   1.637508     1.39   0.165    -.9358174    5.483097
           t |  -.2088293   .1393846    -1.50   0.134    -.4820182    .0643596
       _cons |   14.31443     7.8996     1.81   0.070    -1.168505    29.79736
------------------------------------------------------------------------------
```

FIGURE 6.17
Solution Exercise 6.3.

TABLE 6.3

Exercise 6.4: A Choice of Commute to Work

Alternative	Time (Hours)	Cost ($/Month)	Comfort	Utility
1. Walk	4.5	10	3	5
2. Bike	1.18	10	4	6
3. Bus	1	101	1	5
4. Car	0.42	300	10	7
5. Walk–bus	0.82	101	0	6
6. Car–metro(1)	0.84	145	8	9
7. Car–metro(2)	1	305	9	7
8. Car–train(1)	0.92	175	7	6
9. Car–train(2)	1	175	7	5
10. Walk–train	1.25	125	6	4
11. Car–bus	0.67	151	5	5

```
Multinomial logistic regression                 Number of obs   =         11
                                                LR chi2(12)     =      21.33
                                                Prob > chi2     =     0.0458
Log likelihood = -5.4869943                     Pseudo R2       =     0.6602
```

utility	Coef.	Std. Err.	z	P>|z|	[95% Conf.	Interval]
4						
time	-2.011247	3.763202	-0.53	0.593	-9.386988	5.364493
cost	-.0979995	.1504885	-0.65	0.515	-.3929516	.1969527
comfort	2.761783	6.828313	0.40	0.686	-10.62146	16.14503
_cons	-1.423014	29.57849	-0.05	0.962	-59.39578	56.54975
5	(base outcome)					
6						
time	-5.768677	6.879041	-0.84	0.402	-19.25135	7.713996
cost	-.0796008	.090365	-0.88	0.378	-.2567131	.0975114
comfort	.7578621	1.000833	0.76	0.449	-1.203735	2.719459
_cons	12.69038	13.38978	0.95	0.343	-13.5531	38.93387
7						
time	-5.550071	66629.18	-0.00	1.000	-130596.3	130585.2
cost	.3624063	603.9137	0.00	1.000	-1183.287	1184.012
comfort	-3.05444	22945.48	-0.00	1.000	-44975.37	44969.26
_cons	-56.15375	123583.4	-0.00	1.000	-242275.2	242162.9
9						
time	-5.03359	206006.3	-0.00	1.000	-403770.1	403760
cost	-.5375045	309.1656	-0.00	0.999	-606.491	605.416
comfort	26.58121	24854.14	0.00	0.999	-48686.63	48739.79
_cons	-106.6894	336759.8	-0.00	1.000	-660143.8	659930.4

FIGURE 6.18
Solution Exercise 6.4.

Reference

1. De la Barra, T. 1992. Integrating land use and transport modelling: Decision chains and hierarchies, cambridge urban and architectural studies. ISBN 9780521022170, 196 pages.

7

Transport and Municipal Engineering

7.1 Introduction

This chapter introduces the reader to several classical problems related to transportation systems analysis. Some of these classical problems serve to develop a foundation for more sophisticated problems closer to those faced in real life. Several simplifications serve to facilitate the formulation, and as we advance, more elements will be brought into the model. The problems will take advantage of methods learnt in previous chapters using tools readily available in Excel. More advanced and complex problems may need to be solved elsewhere.

7.2 Location of a Facility within an Area

How can planners at a municipality (or government body) choose the best location for a facility that must provide service to the public at large? You can think of a health clinic, a fire station, ambulance or any other emergency response unit. On a larger scale, you can even think of a regional park, a land field, an airport, an intermodal station, a library or even a market.

Two main indicators are important for this problem: travel time between all pairs of user location and the facility and the overall cost. Often, we engineers must attempt to reduce to a minimum the number of facilities but guarantee a sufficient coverage. For instance, you may want to build a health clinic that anybody can reach in 15 minutes or less or a fire station that can reach help to any location in town in no more than 10 minutes.

Sometimes, this task is impossible, and a single station cannot deliver the expected service on time. In such cases, two or more facilities would be needed. Having two facilities opens a new dimension to this problem; we will call it 'coverage'. Coverage relates to the ability to have as many duplications as possible, that is, we will locate the emergency response unit in such

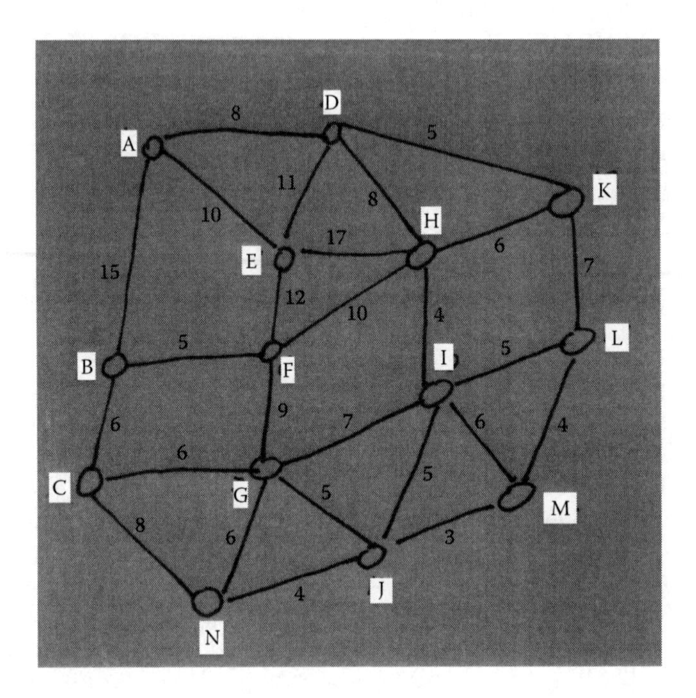

FIGURE 7.1
Handmade node structure.

a way that it maximizes coverage in addition to minimizing total network travel time and cost.

Let's look at the first example. Suppose we have to find the best location for a police station. Presume for the first formulation that one police station is enough, and we will only attempt to minimize total network travel time between the police station and the different neighbourhoods.

A simplified way to solve the problem is by taking a representation of the city based on its road network (links) and its neighbourhoods (nodes). Simply take a piece of paper and prepare a drawing like the one shown in Figure 7.1.

Even though nodes should carry the name of places, we have numbered them for convenience. The links do not necessarily represent actual roads; rather, they serve to connect neighbourhoods. Additionally, links are presumed to serve traffic moving in both directions; in some cases, we could restrict it to one direction of movement, and this would be represented by an arrow.

When assembling this type of problems, the municipal engineer needs to obtain a good estimation of the travel time, which can be initially based on field observations, but he or she must bear in mind that the travel time changes throughout the day, as congestion increases during morning, midday and evening peak hours (when most of the commuting from/to work occurs).

Travel time (in minutes) and possible directional restrictions (arrows) had been added to our example (Figure 7.1). Remember, many other variables could influence the travel time, for instance, surface type (gravel or paved), road condition (potholes will slow you down), alignment adequacy (unnecessarily curvy roads), type of terrain (rolling, mountainous) and intersections. We will further elaborate this in Chapter 8 when a framework to prioritize infrastructure decisions is proposed.

Table 7.1 shows the travel time between any two pairs of cities. This technique is known as origin–destination pairs, and for now we will use it to solve the first version of our problem without the need to use a computer. As you can see, in some instances there is more than one way to move between given origin–destination pairs.

For instance, consider the movement between E and M; you can go through route EFGJM with a duration of 29 minutes, or through route EFIM with a duration of 25 minutes, or through EHIM with a duration of 27 minutes, or through the route EHKLM with a duration of 34 minutes, or through route EHKLIM with a duration of 41 minutes. I will stop here: I could, of course, elaborate more through EABCNJM and many other possible routes.

When solving this type of problem, the decision maker must identify the shortest path between any origin–destination pairs; the shortest path will be defined in terms of the cost, which in this case relates to the travel time.

Final results from all calculations show that node H is the ideal location for the siting of the facility (police station). This conclusion is simply justified because H is the node with the smallest total network travel time of 143 minutes.

TABLE 7.1

Travel Time between Origin–Destination Pairs

O/D	A	B	C	D	E	F	G	H	I	J	K	L	M	N
A	0	15	21	8	10	20	27	16	20	25	13	20	24	29
B	15	0	6	23	17	5	6	15	19	17	21	24	20	14
C	21	6	0	29	23	11	6	15	13	11	19	18	19	8
D	8	23	29	0	11	18	19	8	12	17	5	12	16	37
E	10	17	23	11	0	12	21	17	21	26	23	30	27	27
F	20	5	11	18	12	0	9	10	14	14	16	21	17	15
G	27	6	6	19	21	9	0	11	7	5	17	12	8	6
H	16	15	15	8	17	10	11	0	4	9	6	9	10	13
I	20	19	13	12	21	14	7	4	0	5	10	5	6	9
J	25	17	11	17	26	14	5	9	5	0	14	7	3	4
K	13	21	19	5	23	16	17	6	10	14	0	7	11	18
L	20	24	18	12	30	21	12	9	5	7	7	0	4	11
M	24	20	19	16	27	17	8	10	6	3	11	7	0	7
N	29	14	8	37	27	15	6	13	9	4	18	11	7	0

A simple summation of travel times per node suffices to reveal the location with the higher degree of accessibility where the police station should be located. In a small case, like the one in the aforementioned example, we can do this by inspection, but in larger problems we need to create a mechanism that explicitly models it.

In a nutshell, we are finding the best routes between origin–destination pairs (OD) and then selecting the location that minimizes the total network travel time. More formally, the identification stage should be the result of solving multiple shortest path sub-problems between all possible combinations of pairs of nodes. Hence, the problem has a combinatorial component, which is not conditioned by the decision variables but rather fixed and linked to its very own geometrical structure. Soon we will tackle a somewhat similar problem of municipal infrastructure in which the problem is also combinatorial but the structure does depend on the decision variables.

The second stage of the facility siting problem is that related to the objective. The problem can be extended when a time restriction is imposed. Consider, for instance, that the maximum travel time from any point must be equal to or less than 15 minutes. This restriction will force us to look for a location where all travel times are smaller than the requested one. In this case, the travel times between H and any other city are smaller than or equal to 15 minutes except between H and A and H and E, where the travel time exceeds by 2 minutes and 1 minute, respectively.

7.3 Shortest Path

The shortest path problem is one of the most recognized optimization algorithms; it is used by travel agents, delivery companies and emergency response vehicles even though you use it whether through for a navigation application to drive or plan your commute trip by public transportation or by booking a flight using any of the available online sites on the Internet. The shortest path problem refers to one in which a traveller wants to find the shortest way to move from a given origin to a given destination. By 'shortest' we mean the fastest and, in a general sense, the cheapest. For instance, consider an individual taking the metro to go to work. He or she will find the best way to transfer between metro lines in such a way that minimizes his or her total travel time. Perhaps another user, unfamiliar with the network, may use a rather different definition of 'shortest' as that with fewer connections, or a more straightforward way to get there.

The problem could relate to the travel time, distance or the overall cost. For instance, consider a tourist taking a flight for vacation who wants to get to a given destination at the lowest cost; a similar case is that of the driver taking the highways and other fast roads trying to avoid arterial and collector roads because he or she either wants to minimize the likelihood of accidents (cost) or minimize vehicle deterioration by taking roads in better conditions even if that means a slightly longer travel time. Another case is taking some roads with longer travel time that helps avoid paying tolls.

This problem follows a set-up in which each node is given a decision variable (x), which carries the information of origin (i) and destination (j). An enumeration of all possible movements is performed by utilizing those permitted movements at the constraints. The mathematical algorithm for this problem has one objective: to minimize the total cost ($c_{i,j}$) and three main constraints, namely, one for the origin, one for the destination and a set of constraints for the network in order to define connectivity. The total cost takes the form of a summation over all possible routes. For this, the connectivity of the network must be defined by enumerating all possible permitted movements in the network.

$$\min_{x_{i,j}} \sum_{i=1}^{N} x_{i,j} c_{i,j}$$

The constraints for the starting and ending nodes depend on the number of travellers; in the classical problem there is one individual. The starting node constraint equates all possible outgoing movements to proximal locations (say indexed by b) directly connected to the starting node (say a) to the number of individuals (one in this case), as shown in the following equation:

$$\sum_{b=1}^{B} x_{a,b} = 1$$

For simplicity, consider the case depicted in Figure 7.1. Assume that the starting node is A; hence, the first constraint will take the form

$$x_{A,D} + x_{A,E} + x_{A,B} = 1$$

Consider now M as the ending node (Figure 7.1). Assume again that there is one individual travelling. The constraint for such a node will be $x_{J,M} + x_{I,M} + x_{L,M} = 1$. In general, the constraint for the ending point (Z) for one individual with nearby arrival nodes indexed by r (as you can see, the total value goes to R, which is just a convention since we do not know how many of them are there) is depicted by the following equation:

$$\sum_{r=1}^{R} x_{r,Z} = 1$$

Finally, when there are intermediate nodes, we need to define possible in and out movements from such nodes. This type of constraint is the one that really gives structure to the network. All possible movements should be defined. Note: I say possible because for this you need to take into consideration the permitted direction of the travellers' flow; in most cases, the direction is in both senses, but in some exceptional cases, it may be restricted, or the opposite direction may not make sense at all. For instance, when travelling to another country, it won't make sense to cross the Atlantic and reach Europe to come back to America and then back again to Europe! Assume we are building this for node c and indexing the arrival nodes by a and destination nodes by d. Then the equation for the intermediate nodes will take the form 'all in = all out', as shown as follows:

$$\sum_{a=1}^{A} x_{a,c} = \sum_{d=1}^{D} x_{c,d}$$

It is easier to see this in an example; so let's look at one concrete case you may be familiar with.

7.3.1 Simple Example of the Shortest Path

Consider the case of a traveller choosing his or her connections to fly from Montreal (YUL) to Dubai (DBX). Assume the individual prices of flight segments are known, as well as possible connections through London (LHR), Paris (CDG) or Zurich (ZRH). The network of stopovers is given in Figure 7.2, and the complete mathematical formulation is also shown below it.

As you probably realize, I have used airport codes instead of the full airport names. This is because, even without using the actual names, the mathematical formulation turns out to be too long, and hence it is advisable to use just one letter per node to minimize this issue. This letter can represent the actual name or it can be arbitrarily chosen.

The other thing you probably realize is that this problem is somewhat trivial once you know the cost. But don't jump into a conclusion too fast because defining the cost could be somewhat challenging. You probably remember similar cases in which I combined various criteria into one in the previous chapters. For this case, the cost could comprise the actual monetary cost combined with flight comfort (first, business or economy seat?), the total travel time and whether the flights are overnight or not (to help you adjust to jet-lag), stopover convenience (perhaps you prefer a flight stopping for hours to be able to see a new city!), arrival and departure time (some people may not care for this) or whether you make air miles (very important to me, I actually use this as an initial filter).

Let's say I weighed all these criteria, as shown in a previous chapter, and the final cost is given in Figure 7.3. Hence, it is simple to see that the solution involves only two nodes with values of 1, $x_{yul,zrh} = 1, x_{zrh,dbx} = 1$, and all

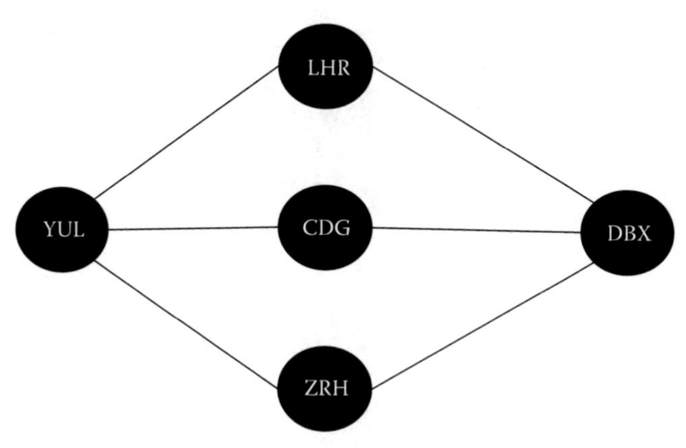

$$\min_{x_{i,j}} (x_{yul,lhr}c_{yul,lhr} + x_{yul,cdg}c_{yul,cdg} + x_{yul,zrh}c_{yul,zrh} + x_{lhr,dbx}c_{lhr,dbx} + x_{cdg,dbx}c_{cdg,dbx} + x_{zrh,dbx}c_{zrh,dbx})$$

$$x_{yul,lhr} + x_{yul,cdg} + x_{yul,zrh} = 1 \; x_{lhr,dbx} + x_{cdg,dbx} + x_{zrh,dbx} = 1$$

FIGURE 7.2
One-scale flight network.

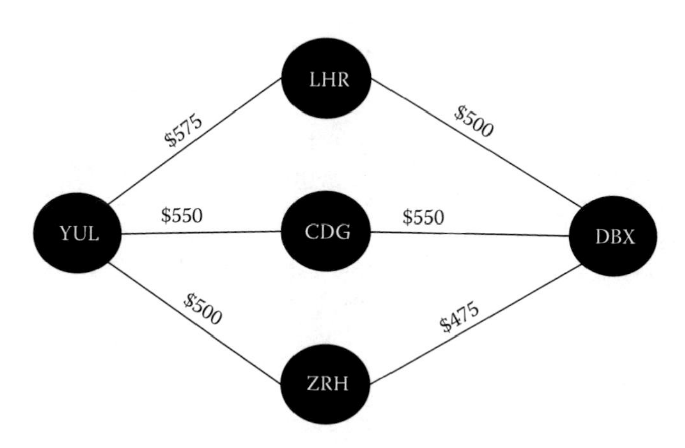

FIGURE 7.3
One-scale flight network with costs.

others with values 0, $x_{yul,lhr} = 0, x_{yul,cdg} = 0, x_{cdg,dbx} = 0, x_{lhr,dbx} = 0$. The objective value would be 975.

7.3.2 Extended Example of the Shortest Path

Let's now extend the example. Consider you just have discovered that there are other routes with more stops (two each) in addition to the previous routes. Figure 7.4 illustrates this expanded network used in this section.

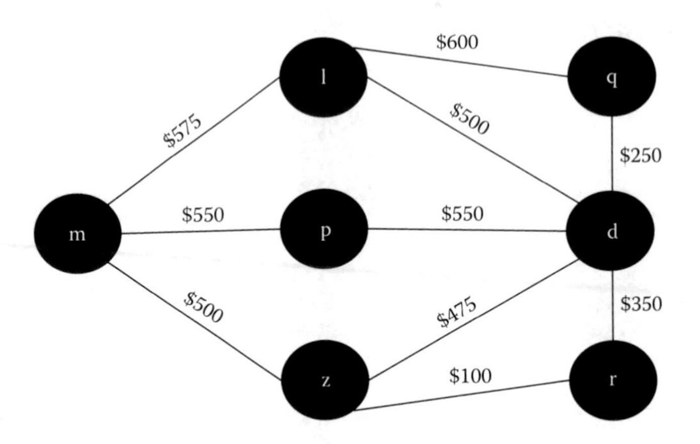

FIGURE 7.4
An expanded flight network.

The first thing we will do is change the airport codes to single letters; so Montreal (m), London (l), Paris (p), Zurich (z), Qatar (q), Dubai (d) and Rome (r) would be our possible choices.

In this formulation, we need to make use of constraints for the starting and ending nodes as well as for the intermediate nodes (l and z when connecting to q and r).

So far, this problem looks like the previous one; so to make it more interesting, let us say there are two travellers (you and your friend). The nodes that take values would need to take a value of either 2 or 0 because you want to be in the same plane as your very dear friend (won't you?). This will require the inclusion of additional constraints stipulating that the nodes take the values 0 or 2. In the case of various unrelated travellers, we do not need to impose such a constraint.

The objective takes the same form; the only difference is that there are more nodes, and the cost for the new possible movements needs to be defined a priori (in advance). The constraints also take the well-known form; the departure node constraint is the same (except that the number of travellers is equal to two). The arrival constraint has more possibilities by adding the combinations Qatar (q) and Rome (r).

$$\min_{x_{i,j}}(x_{m,l}c_{m,l} + x_{m,p}c_{m,p} + x_{m,z}c_{m,z} + x_{l,d}c_{l,d} + x_{p,d}c_{p,d} + x_{z,d}c_{z,d} + x_{l,q}c_{l,q}$$

$$+ x_{z,r}c_{z,r} + x_{r,d}c_{r,d} + x_{q,d}c_{q,d})$$

$$x_{m,l} + x_{m,p} + x_{m,z} = 2$$

$$x_{l,d} + x_{p,d} + x_{z,d} + x_{q,d} + x_{r,d} = 2$$

$$x_{i,j} = (0, 2)\forall i, j$$

We also need to define the intermediate node constraints; they will take the following form: $x_{m,l} = x_{l,d} + x_{l,q}$ and $x_{m,z} = x_{z,d} + x_{z,r}$. Note we have not used the possible connections between London and Paris and Paris and Zurich. I will leave this to you as home practice. Simply add an expression for the cost and proceed in the same manner as before.

One thing that probably crossed your mind is why we have to write down all combinations at the objective even though some would not be used. The answer lies in the fact that you want to define all possibilities first, and then you choose the most convenient one (typically the cheapest); the circumstances that define the most convenient one may change along the way, and then you may find yourself revisiting the problem. So if the structure is in place, you can deal with any changes and will be able to make decisions even under changing circumstances.

This is typical of optimization problems, where we run scenarios to answer specific questions and examine alternative solutions. We will see a more detailed case of scenarios in the following chapter, but it is applicable anywhere. If you change from one decision maker to another, the new individual's definition and weighing of the cost will likely be different from those of the first one.

The shortest path is used in many applications in real life: websites of travel agencies and airlines, navigation systems, public transportation applications, etc.

7.4 Transportation (Distribution or Allocation)

The transportation problem is nothing but the identification of supply and demand forces from origins (O) to destinations (D). The purpose is to identify the amounts that will go between *OD* pairs. Think of a company that manufactures the same product at several factories across Asia and ships the goods out of seaports to locations where they are demanded. Picture, for instance, computers produced in Hong Kong, Singapore and Shanghai heading to Los Angeles, New York and Panama. How the computer goes from the factory to the port is itself another problem, which we will address soon. The transportation problem does not worry about roads or routes but rather concentrates on finding the cheapest way to move the goods from the origin to the destination. The fact that a factory is more competitive and can afford to sell at the lowest prices is central to this problem; goods out of that factory will be moved first. But eventually it will run out of commodities, and the second cheapest should provide goods until it reaches its capacity, then the third one, and so on. To make it simple, the supply and demand will be assumed to be at equilibrium, that is, the number of goods supplied matches with the demand. Then this assumption will be relaxed and the supply will be much larger than the demand. Demand in that event will be known.

Costs are associated with origin–destination pairs, which assumes that in each pair the cheapest route had been identified. How the route goes is not relevant in the scope of this problem; it all comes down to a final cost per route.

The solution takes the form of the number of units shipped between the origin and the destination, that is, how many units are moved from Hong Kong to New York, from Hong Kong to Los Angeles, from Hong Kong to Panama, then from Singapore to each of the three destinations, and from Rotterdam to each of them again. At the end, for this case you have nine possible combinations of origin–destination pairs, along with the number of units moved between them.

The algorithm behind this type of problem takes the form of one objective and two major constraints. The objective is the common minimization of the total cost, with the decision variable being the number (x) moved from the origin (i) to the destination (j). The total cost is found by multiplying individual cost ($c_{i,j}$) with the corresponding amounts ($x_{i,j}$).

$$\min_{x_{i,j}} \sum_{i=1}^{N} x_{i,j} c_{i,j}$$

The first constraint deals with the equilibrium of the numbers exported out of a given origin: it makes the summation of all goods out of that origin to any possible destination (D_i) equal to the total available supply at such origin. The second constraint does the same but for a given destination, that is, it sums across origins (O_i) to a common destination.

$$\sum_{j=1}^{N} x_{i,j} = O_i, \quad \text{for } i = 1, 2, 3...$$

$$\sum_{i=1}^{N} x_{i,j} = D_j, \quad \text{for } j = 1, 2, 3...$$

Lack of equilibrium results in a slight modification on the previous equations: goods available at a given origin (O_i for a given i) may not have been totally utilized, hence the 'smaller than' inequality. This is certainly true when goods are more expensive and hence left to completely unsatisfied demand at a given destination. The equation for demand remains an equality if we enforce the assumption that demand will be satisfied; on the contrary, the equality will turn out to be a 'bigger than' inequality if the demand (D_j for a given j) is not satisfied or suppliers send more goods than possibly required to a given destination (remember containers will force you to round up your number of shipped units, and demand is difficult to estimate, so you may actually ship more units at the end of the day).

$$\sum_{j=1}^{N} x_{i,j} \leq O_i, \quad \text{for } i = 1,2,3\ldots$$

$$\sum_{i=1}^{N} x_{i,j} \geq D_j, \quad \text{for } j = 1,2,3\ldots$$

In addition to the previous constrains, the number of goods needs to be positive ($x_{i,j} \geq 0$) depending on the type of goods being moved, and it could be restricted to be integers ($x_{i,j} \in Z$), strictly speaking for all pairs of i, j.

7.4.1 Simple Example of Transportation

Let's take the case of mobile phones produced in Hong Kong (k), Shanghai (s) and Beijing (b) and shipped to the United States for sale in New York (n), Houston (h) and Los Angeles (l). You can think of them as coming from the same company (whichever you prefer). The objective is to minimize the total cost:

$$\min_{x_{i,j}} \sum_{i=1}^{N} x_{i,j} c_{i,j}$$

In this case, we can explicitly write it for the given nodes, so it takes the form

$$\min_{x_{i,j}} x_{k,n}c_{k,n} + x_{k,h}c_{k,h} + x_{k,l}c_{k,l} + x_{s,n}c_{s,n} + x_{s,h}c_{s,h} + x_{s,l}c_{s,l} + x_{b,n}c_{b,n} + x_{b,h}c_{b,h} + x_{b,l}c_{b,l}$$

For this case, we have six constraints for the origins (supply–manufactures) and six additional constraints for the destinations (demand–locations). Assume that the corresponding capital letters for each city represent the number being produced or demanded at each one of them; that is, K, S and B represent the production out of Hong Kong, Shanghai and Beijing, respectively. For New York, Houston and Los Angeles, we will use N, H and L to represent the numbers being demanded. For instance, $x_{k,n}$ means the number produced at k being used to satisfy the demand at n. Once again, the inequality is smaller or equal, given the fact that not necessarily all production will sell.

$$x_{k,n} + x_{k,h} + x_{k,l} \leq K$$

$$x_{s,n} + x_{s,h} + x_{s,l} \leq S$$

$$x_{b,n} + x_{b,h} + x_{b,l} \leq B$$

For the destinations, the constraints look much like those of the origins, except in the sense of the inequality. As expected, the inequality is greater or equal, given the fact that the demand is an estimation and you don't want to fall short of supplying at least as much as required and perhaps some extra units, just in case.

$$x_{k,n} + x_{s,n} + x_{b,n} \geq N$$
$$x_{k,h} + x_{s,h} + x_{b,h} \geq H$$
$$x_{k,l} + x_{s,l} + x_{b,l} \geq L$$

Again, just as in the case of the shortest path, we arrive at the point at which the cost is key to the solution. Once again, the cost in this case relates to many elements, such as the carrier's fees and the total travel time. The solution will identify the number of goods from given sources to supply to the destinations, but will not be concerned with the selected route. Route selection leads again to the shortest path problem.

Let's assume that the costs (Table 7.2) are provided to you by the analyst who has compared air transportation versus sea transportation and many possible routes and carriers and who finally identified the most convenient route for each pair (origin–destination).

We will illustrate the use of Excel in Figure 7.5. First, the problem needs to be defined in such a way that the structure is represented explicitly and all elements whose value may change are defined in terms of the other elements with fixed or constant values. Let's see; first, let's place on a spreadsheet the coefficients next to each variable in order to define the constraints. Enumeration of pairs of origin–destination is defined: *kn*, *kh*, *kl*, *sn*, *sh*, *sl*, *bn*, *bh*, *bl*. Then, they are also defined as variables in another region of the spreadsheet. To define the first constraint, we multiply each coefficient by the variable, so, for instance, the first constraint only has coefficients of 1 for those elements present at it (kn, kh, kl); the rest carry a zero next to them (sn, sh, sl, bn, bh, bl).

The multiplication continues for all remaining coefficients with the corresponding variables. This exhaustive multiplication facilitates the definition of all remaining constraints by simply copying it to the other ones as shown in Figure 7.6.

TABLE 7.2

Origin–Destination Costs

Origin/Destination	New York (n)	Houston (h)	Los Angeles (l)
Hong Kong (k)	1400	1200	900
Shanghai (s)	1300	1100	800
Beijing (b)	1200	1500	1000

Constraints	kn	kh	kl	sn	sh	sl	bn	bh	bl	total	limiting
1	1	1	1							+G2*G11+H2*G12+I2*G13	
2				1	1	1					
3							1	1	1		
4	1			1			1				
5		1			1			1			
6			1			1			1		

Variables	
kn	0
kh	0
kl	0
sn	0
sh	0
sl	0
bn	0
bh	0
bl	0

FIGURE 7.5
Problem definition.

Constraints	kn	kh	kl	sn	sh	sl	bn	bh	bl	total	limiting
1	1	1	1							0	
2				1	1	1				0	
3							1	1	1	0	
4	1			1			1			0	
5		1			1			1		0	
6			1			1			1	0	

FIGURE 7.6
Constraint set-up.

Next, we need to define the objective. For this purpose, we add the cost of each pair and multiply the binary decision variable by the cost, as shown in Figure 7.7.

The objective is the overall summation of the cost for all links, as shown in Figure 7.8. For those selected, the decision variable will take the number of units assigned; for the others it will be zero.

The next step is to bring the limiting values next to each constraint, as shown in Figure 7.9, and invoke the solver tool located at the DATA tab of Excel.

Variables	SOL	Cost	
kn	0	1400	+G10*H10
kh	0	1200	

FIGURE 7.7
Objective set-up for one variable.

Variables	SOL	Cost	
kn	0	1400	0
kh	0	1200	0
kl	0	900	0
sn	0	1300	0
sh	0	1100	0
sl	0	800	0
bn	0	1200	0
bh	0	1500	0
bl	0	1000	0
			=SUM(I10:I18)
			SUM(**number1**, [number2], ...)

FIGURE 7.8
Objective set-up across variables.

Constraints	kn	kh	kl	sn	sh	sl	bn	bh	bl	total	limiting	
1	1	1	1							0	**500**	**K**
2				1	1	1				0	**400**	**S**
3							1	1	1	0	**100**	**B**
4	1			1			1			0	**300**	**N**
5		1			1			1		0	**250**	**H**
6			1			1			1	0	**450**	**L**

FIGURE 7.9
Limiting values on the constraints.

We invoke the solver and select the objective cell (with the summation of cost) and the constraints. The constraints require the definition of the left-hand side and the right-hand side, as shown in Figure 7.10.

All other constraints should also be defined. After that, one needs to specify the decision variables. The final problem set-up on Excel is shown in Figure 7.11.

The final solution is then identified by clicking on solve. This populates the cells corresponding to the decision variables with the optimal values, as shown in Figure 7.12.

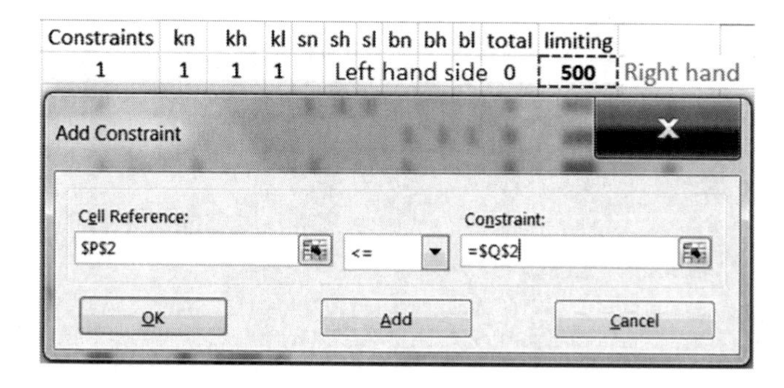

FIGURE 7.10
Solver: definition of the left-hand side.

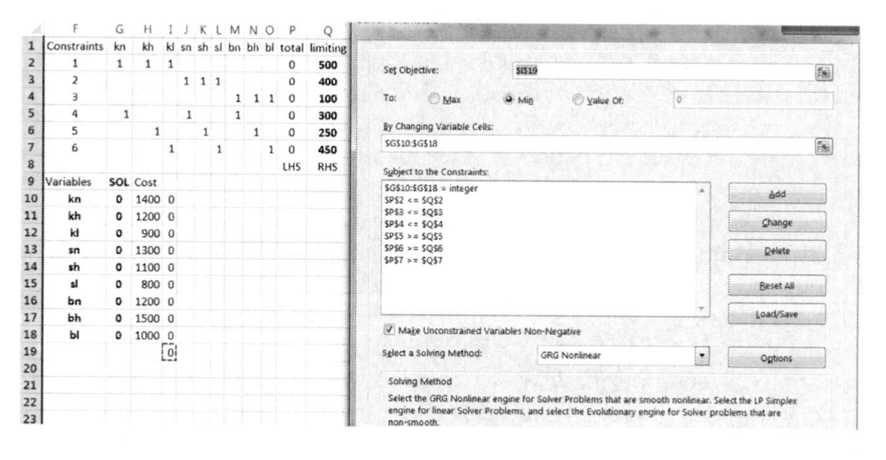

FIGURE 7.11
Final problem set-up in Excel.

Variables	SOL	Cost	
kn	200	1400	280000
kh	176	1200	211200
kl	124	900	111600
sn	0	1300	0
sh	74	1100	81400
sl	326	800	260800
bn	100	1200	120000
bh	0	1500	0
bl	0	1000	0
			1065000

FIGURE 7.12
Optimal values for the decision variables.

7.5 Trans-Shipment

The trans-shipment problem looks much like the shortest path, but it is really an extended version of it. Trans-shipment allows flows of commodities coming into some nodes: you can think of them as the location of airports, ports, or simply entry points, that is, locations where flows from outside the system come into it. Flows exiting the network are also permitted. We typically use arrows pointing outwards to represent exiting movements and arrows pointing inwards to denote income flows. In trans-shipment, we could find several links being used to move goods.

The constraints for this type of problem follows one general rule already used in the shortest path: all incoming flows equal all outgoing flows. However, among incoming flows one can have interior ($x_{i,j}$) and exterior flows coming in (I), and among outgoing flows they can head towards an interior ($x_{i,j}$) or exterior location (E). The following is the generic constraint used for this type of problem; soon we will see its application in a couple of true examples. Interior flows coming in are indexed by i, whereas those heading out but into interior nodes are indexed by j.

$$\sum_{i=1}^{N} x_{i,j} + I = \sum_{j=1}^{M} x_{i,j} + E$$

The objective is just a replica of the shortest path objective and the trans-shipment objective, that is, to minimize total cost given by a multiplication of unitary cost by the number of units on the link, which is shown here:

$$\min_{x_{i,j}} \sum_{i=1}^{n} \sum_{j=1}^{m} x_{i,j} c_{i,j}$$

7.5.1 Example of Trans-Shipment for Canadian Airports and Roads

This example illustrates an application of the trans-shipment algorithm (Figure 7.13). Assume you have international passengers arriving at Canadian airports in Vancouver (V), Regina (R), Toronto (T), Montreal (M) and Halifax (H). Passengers rent cars and head to their favourite destinations; for simplicity we will exclude those passengers taking local flights and assume that all individuals in our network have chosen to move by highways. The highway network is shown in Figure 7.13; the numbers over the arrows represent cost (in this case distance in kilometres), the grey arrows are the flows of international passengers in hundreds of thousands, and the black arrows represent outgoing flows exiting the system at specific cities. The codes for cities are Vancouver (V), Calgary (C), Jasper (J), Edmonton (E), Regina (R), Saskatoon (S), Winnipeg (W), Thunder Bay (B), Toronto (T), Ottawa (O), Montreal (M), Quebec (Q), Saint Andrews (A), Charlottetown (C) and Halifax (H).

The set-up of constraints is given in Table 7.3; it is convenient as you will learn after this chapter to write the constraints before coding your software

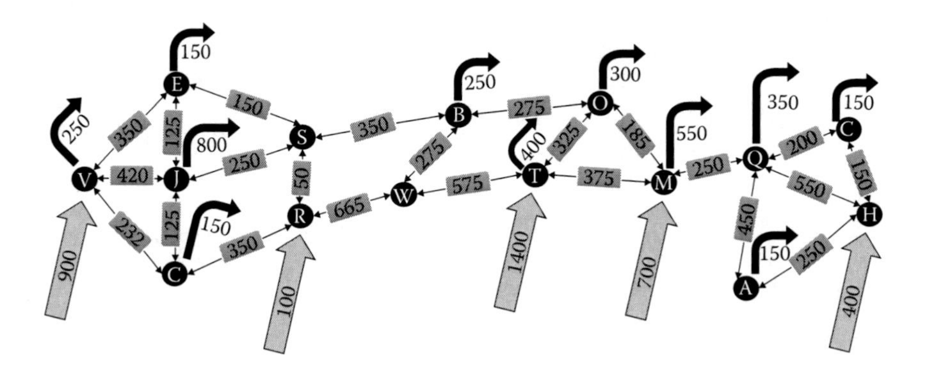

FIGURE 7.13
Canadian network.

TABLE 7.3

Definition of Constraints for Canadian Network

Node	Constraint
V	$900 + x_{E,V} + x_{J,V} + x_{C,V} = 250 + x_{V,E} + x_{V,J} + x_{V,C}$
E	$x_{V,E} + x_{J,E} + x_{S,E} = 150 + x_{E,V} + x_{E,J} + x_{E,S}$
J	$x_{V,J} + x_{E,J} + x_{C,J} + x_{S,J} = 800 + x_{J,V} + x_{J,E} + x_{J,C} + x_{J,S}$
C	$x_{V,C} + x_{J,C} + x_{R,C} = 150 + x_{C,V} + x_{C,J} + x_{C,R}$
S	$x_{E,S} + x_{J,S} + x_{R,S} + x_{B,S} = x_{S,E} + x_{S,J} + x_{S,R} + x_{S,B}$
R	$100 + x_{C,R} + x_{S,R} + x_{W,R} = x_{R,C} + x_{R,S} + x_{R,W}$
B	$x_{S,B} + x_{W,B} + x_{O,B} = 250 + x_{B,S} + x_{B,W} + x_{B,O}$
W	$x_{R,W} + x_{B,W} + x_{T,W} = x_{W,R} + x_{W,B} + x_{W,T}$
O	$X_{B,O} + x_{T,O} + x_{M,O} = 300 + X_{O,B} + x_{O,T} + x_{O,M}$
T	$1400 + x_{W,T} + x_{O,T} + x_{M,T} = 400 + x_{T,W} + x_{T,O} + x_{T,M}$
M	$700 + x_{T,M} + x_{O,M} + x_{Q,M} = 550 + x_{M,T} + x_{M,O} + x_{M,Q}$
Q	$x_{M,Q} + x_{A,Q} + x_{C,Q} + x_{H,Q} = 350 + x_{Q,M} + x_{Q,C} + x_{Q,H} + x_{Q,A}$
A	$x_{Q,A} + x_{H,A} = 150 + x_{A,Q} + x_{A,H}$
C	$x_{Q,C} + x_{H,C} = 150 + x_{C,Q} + x_{C,H}$
H	$400 + x_{Q,H} + x_{C,H} + x_{A,H} = x_{H,Q} + x_{H,C} + x_{H,A}$

(Excel in this case). For the trans-shipment problem, we require one constraint per node, so in total we will have 15 constraints as follows. For the formulation on Excel, we need to move all variables to one side and amounts to the other side; I have chosen to move all variables to the left-hand side and amounts to the right-hand side. As you can see in Figure 7.14, we are looking at 46 origin–destination pairs.

Figure 7.15 shows the solution for the network: from Vancouver, 25 travellers will go to Jasper and 625 to Calgary, from Saskatoon to Edmonton 150, from Calgary to Jasper 475, from Regina to Saskatoon 100, Thunder Bay and Saskatoon 350, Ottawa and Thunder Bay 600, Toronto and Ottawa 900, Toronto and Montreal 100, Montreal and Quebec 250, Charlottetown and Quebec 100, Halifax and Saint Andrews 150 and Halifax and Charlottetown 250. The rest of the nodes receive no travellers.

7.6 Allocation of Public Works

The optimization of allocations is nothing but an application of the transportation (distribution) problem in the world of public affairs, to be precise in the tendering process for public infrastructure construction or maintenance. Think of a municipality or any agency trying to allocate public works among

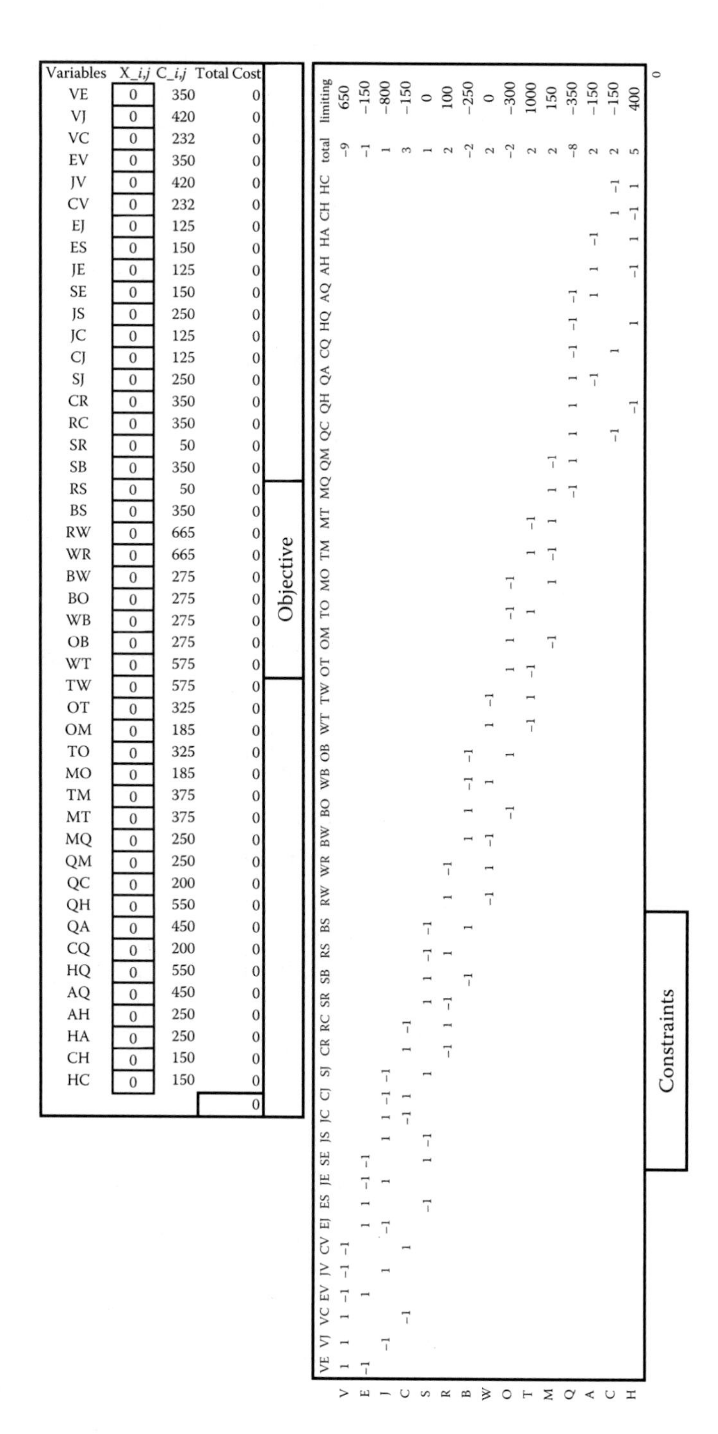

Variables	X_i,j	C_i,j	Total Cost
VE	0	350	0
VJ	0	420	0
VC	0	232	0
EV	0	350	0
JV	0	420	0
CV	0	232	0
EJ	0	125	0
ES	0	150	0
JE	0	125	0
SE	0	150	0
JS	0	250	0
JC	0	125	0
CJ	0	125	0
SJ	0	250	0
CR	0	350	0
RC	0	350	0
SR	0	50	0
SB	0	350	0
RS	0	50	0
BS	0	350	0
RW	0	665	0
WR	0	665	0
BW	0	275	0
BO	0	275	0
WB	0	275	0
OB	0	275	0
WT	0	575	0
TW	0	575	0
OT	0	325	0
OM	0	185	0
TO	0	325	0
MO	0	185	0
TM	0	375	0
MT	0	375	0
MQ	0	250	0
QM	0	250	0
QC	0	200	0
QH	0	550	0
QA	0	450	0
CQ	0	200	0
HQ	0	550	0
AQ	0	450	0
AH	0	250	0
HA	0	250	0
CH	0	150	0
HC	0	150	0
			0

FIGURE 7.14
Objective and constraints for Canadian network.

Variables	X_i,j	C_i,j	Total Cost	
VE	0	350	0	
VJ	25	420	10,500	
VC	625	232	145,000	
EV	0	350	0	
JV	0	420	0	
CV	0	232	0	
EJ	0	125	0	
ES	0	150	0	
JE	0	125	0	
SE	150	150	22,500	
JS	0	250	0	
JC	0	125	0	
CJ	475	125	59,375	
SJ	300	250	75,000	
CR	0	350	0	
RC	0	350	0	
SR	0	50	0	
SB	0	350	0	
RS	100	50	5,000	Objective
BS	350	350	122,500	
RW	0	665	0	
WR	0	665	0	
BW	0	275	0	
BO	0	275	0	
WB	0	275	0	
OB	600	275	165,000	
WT	0	575	0	
TW	0	575	0	
OT	0	325	0	
OM	0	185	0	
TO	900	325	292,500	
MO	0	185	0	
TM	100	375	37,500	
MT	0	375	0	
MQ	250	250	62,500	
QM	0	250	0	
QC	0	200	0	
QH	0	550	0	
QA	0	450	0	
CQ	100	200	20,000	
HQ	0	550	0	
AQ	0	450	0	
AH	0	250	0	
HA	150	250	37,500	
CH	0	150	0	
HC	250	150	37,500	
			1,092,375	

FIGURE 7.15
Solution for Canadian network.

contractors. In the language of the transportation problem, each contractor will represent a *source* and each region in town a *destination*. The most common way to look at this problem is when a country wants to award public infrastructure maintenance contracts for the coming year. Regions in a country or town are typically chosen to match corridors of interest or zones that are geographically separated from others.

The objective is to minimize total cost, but let me stop you right there. Cost in this case, as in many others, is not only the plain monetary figure that the bidder wrote in his or her bid: no! it should comprise other elements such as the bidder technical capability to undertake the task, and perhaps also their history of previous works (experience) and their financial capability. In some tendering processes, this is separated as a first stage in which an initial filter based on financial and technical capabilities is inserted to reduce the number of bidders so that in the second stage only those that had met with the aforementioned criteria are considered. For the rest of this section, we will assume you have already incorporated the experience together with the cost. Here are two ideas of how you can do this: through a weighted expression, in which you can (a) monetize experience and subtract it from cost or (b) bring both elements to a non-monetized base, by considering the likelihood of successful achievement of the project in a 0–10 scale and then adding to it the number of similar scale projects up to 10.

For us, the capability will be explicitly considered in the formulation, and the use of an initial filter can rule out risky/inexperienced tenders. Let's get the formulation: consider a municipality tendering public works contracts containing packages of roads, sewage and water-main maintenance and repairs (this would be prepared by a coordination approach, which I will explain in the next section). The sources are the contractors with limited construction capability (amount of goods), and the destinations are the urban zones or regions within a municipality in which annual construction works (for infrastructure) need to be done and awarded (allocated) to a contractor. The objective is to distribute the construction projects for all areas to the lowest bidders. Assume we have m contractors in total, indexed by i, and n regions, indexed by j. The contractor's maximum construction capacity is given by a_i and the total amount of work required at a given region by b_j. In this problem, $c_{i,j}$ denotes the unitary cost of construction, which could be for maintenance, rehabilitation or upgrading per metre of pipe or square meter of road. This cost is unique for a given contractor i on a given region j. Finally, let $y_{i,j}$ denote the total amount of work awarded to contractor i on region j, which is not necessarily equal to his or her capacity (a_i). In this context, the problem takes the following objective:

$$\min_{y_{i,j}} \sum_{i=1}^{n} \sum_{j=1}^{m} y_{i,j} c_{i,j}$$

This will be subject to two constraints. The first one is for the contractor's capacity, which cannot be exceeded:

$$\sum_{i=1}^{m} y_{i,j} \leq a_i$$

The other constraint is for the amounts required at a given region (or zone). The total amount of work allocated to contractors in region j cannot exceed the total amount of work b_j required in that zone.

$$\sum_{j=1}^{n} y_{i,j} \geq b_i$$

This problem can be formulated in a reverse way, that is, considering contractors to be destinations and regions/zones to be sources. Disregarding the approach, the mechanism is a simple matching process in which we identify the amounts allocated to origin–destination pairs. Figure 7.16 illustrates such a concept for five contractors, three regions and four types of interventions. As seen, interventions in different zones of the town are allocated based on each contractor's area of expertise and the quoted cost.

Depending on their location, some contractors will be cheaper than others. One mainstream belief is that local contractors would have less expensive

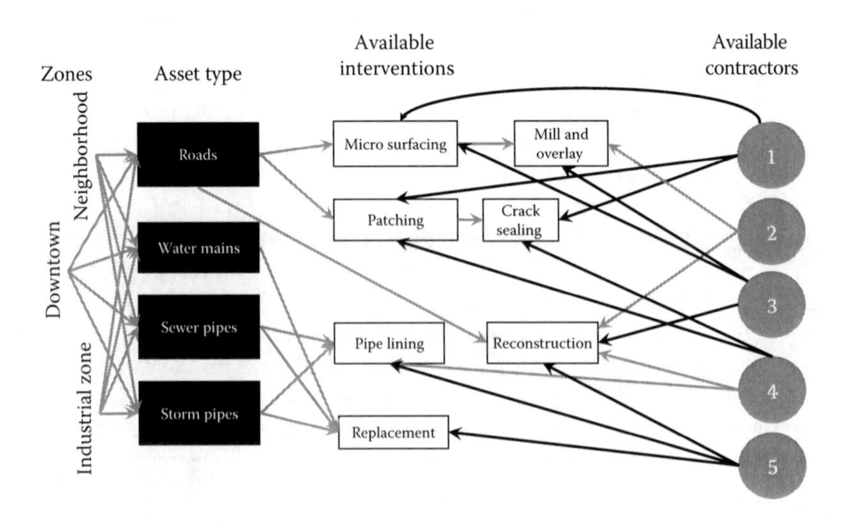

FIGURE 7.16
Allocation of works to contractors.

bids compared to those based at far locations (external), mostly because they have temporary relocation cost added to their overall cost structure when bidding for a project. However, other elements such as technology (machine productivity) may make some contractors (even if not locals) more competitive above certain levels of scale (amount of works). Hence, putting together a good package per zone and awarding it in such a way is sometimes preferable to tendering individual repairs per road segment. Section 7.7 explains the basics on how to identify which public works to bundle together by employing a coordination of works approach. The following section contains an example that hopefully will illustrate the applicability of awarding public works.

7.6.1 Example of Public Works Awarding

Consider the case of five contractors: let's assume we are talking about one job only, namely, pipe replacement, and that the cost shown in Table 7.4 is the unitary cost per linear meter. The regions are downtown (d), highway frontage (h), industrial park (p) and Rosedale residential (r) (Table 7.1).

The rest of the problem concentrates on identifying the amounts allocated to contractor–region pairs. It is useful, however, to explicitly write the constraints before we transfer them into an Excel spreadsheet; let's see for contractors whether we may fully utilize their capability a_i or not. Let's assume the contractors' total capabilities are $1 = 380$, $2 = 290$, $4 = 225$, $3 = 125$, $5 = 150$.

$$x_{1,d} + x_{1,h} + x_{1,p} + x_{1,r} \leq a_1 = 380$$

$$x_{2,d} + x_{2,h} + x_{2,p} + x_{2,r} \leq a_2 = 290$$

$$x_{3,d} + x_{3,h} + x_{3,p} + x_{3,r} \leq a_3 = 225$$

$$x_{4,d} + x_{4,h} + x_{4,p} + x_{4,r} \leq a_4 = 125$$

$$x_{5,d} + x_{5,h} + x_{5,p} + x_{5,r} \leq a_5 = 250$$

TABLE 7.4

Contractors and Their Capabilities

Contractor/Region	d	h	p	r
1	1400	1200	900	1050
2	1300	1100	800	950
3	1200	1500	1000	850
4	900	700	600	1150
5	1700	500	400	1250

For regions, we want to make sure that we allocate all the works that are required this year (b_j); this is, of course, an estimation of works, and along the way, there may be changes mostly in the direction of increasing needs with additional areas to be paved/repaired (remember, though you observe the condition and make the planning, it takes 1 year before you actually do the construction work, so things may have gotten a bit worse). Let's assume the municipality has a budget of \$1 M for this work and that the following amounts of work are being demanded per region: $d = 125$, $h = 150$, $p = 550$, $r = 325$.

$$x_{1,d} + x_{2,d} + x_{3,d} + x_{4,d} + x_{5,d} \geq b_d = 125$$

$$x_{1,h} + x_{2,h} + x_{3,h} + x_{4,h} + x_{5,h} \geq b_h = 150$$

$$x_{1,p} + x_{2,p} + x_{3,p} + x_{4,p} + x_{5,p} \geq b_p = 550$$

$$x_{1,r} + x_{2,r} + x_{3,r} + x_{4,r} + x_{5,r} \geq b_r = 325$$

Figure 7.17 shows the final set-up in Excel. As you can see, I have transferred the constraints directly into the right box and the objective into the left box. Also, remember that each constraint is the product of a coefficient on the constraint table and the value of $X_{i,j}$. As you can notice, the definition of constraints follows the same order of the constraint equations previously shown.

SUM | fx =+F2*B2+G2*B3+H2*B4+I2*B5+J2*B6+K2*B7+L2*B8+M2*B9

	A	B	C	D
1	Variables	X_i,j	C_i,j	Tot.Cost
2	**1d**	**0**	1400	0
3	**1h**	**0**	1200	0
4	**1p**	**0**	900	0
5	**1r**	**0**	1050	0
6	**2d**	**0**	1300	0
7	**2h**	**0**	1100	0
8	**2p**	**0**	800	0
9	**2r**	**0**	950	0
10	**3d**	**0**	1200	0
11	**3h**	**0**	1500	0
12	**3p**	**0**	1000	0
13	**3r**	**0**	850	0
14	**4d**	**0**	900	0
15	**4h**	**0**	700	0
16	**4p**	**0**	600	0
17	**4r**	**0**	1150	0
18	**5d**	**0**	1700	0
19	**5h**	**0**	500	0
20	**5p**	**0**	400	0
21	**5r**	**0**	1250	0
22				0

OBJECTIVE

	F	G	H	I	J	K	L	M	N	O	P	Q	R	S	T	U	V	W	X	Y	Z	AA
1	1d	1h	1p	1r	2d	2h	2p	2r	3d	3h	3p	3r	4d	4h	4p	4r	5d	5h	5p	5r	total	limiting
2	1	1	1	1																	=+F2*	**380**
3					1	1	1	1													0	**290**
4									1	1	1	1									0	**225**
5													1	1	1	1					0	**125**
6																	1	1	1	1	0	**250**
7	1				1				1				1				1				0	**125**
8		1				1				1				1				1			0	**150**
9			1				1				1				1				1		0	**550**
10				1				1				1				1				1	0	**325**

CONSTRAINTS

FIGURE 7.17
Excel set-up for contractor allocation.

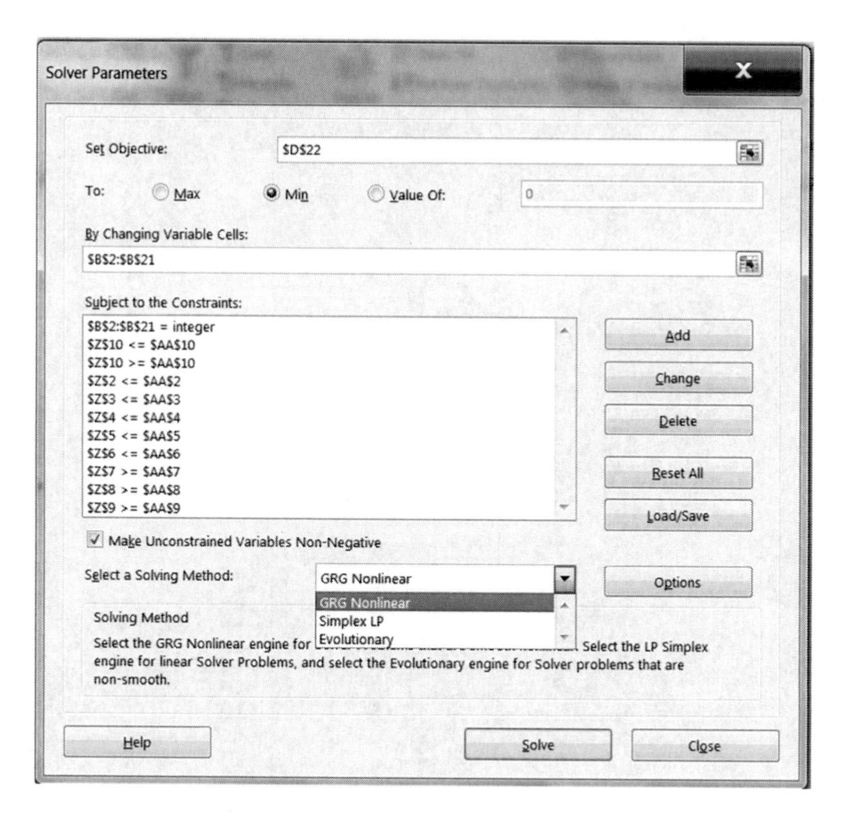

FIGURE 7.18
Solver set-up for contractor allocation.

The set-up in the solver window is shown in Figure 7.18. As you can see, all constraints are given in rounded units (integers); the objective corresponds to the cell that minimizes total cost, and the decision variables match the cells that we need to estimate.

The final solution is shown in Figure 7.19. As can be seen, the industrial park works will be allocated to three contractors (1, 2 and 5), the works on Rosedale residential subdivision to two contractors (2 and 3), the works on downtown exclusively to contractor 4 and the works to highway frontage to one contractor (5).

The problem can be generalized to a more real setting. Consider the case of many types of interventions: say, pipe lining and pipe replacement along with road reconstruction and crack sealing, just to name a few. It could be the case that contractor's expertise is unique to one type of infrastructure, but some large contractors do have expertise across the board of public infrastructure assets. I will leave the formulation of this problem as an home exercise; however, as hints, consider that you need to create one spreadsheet per intervention. So in the case, we just spell out, and you will need four spreadsheets.

	A	B	C	D	E	F
1	Variables	X_i,j	C_i,j	Tot.Cost		
2	1d	0	1400	0		
3	1h	0	1200	0		
4	1p	260	900	234000		
5	1r	0	1050	0		
6	2d	0	1300	0		
7	2h	0	1100	0		
8	2p	190	800	152000		
9	2r	100	950	95000		
10	3d	0	1200	0		
11	3h	0	1500	0		
12	3p	0	1000	0		
13	3r	225	850	191250		
14	4d	125	900	112500	OBJECTIVE	
15	4h	0	700	0		
16	4p	0	600	0		
17	4r	0	1150	0		
18	5d	0	1700	0		
19	5h	150	500	75000		
20	5p	100	400	40000		
21	5r	0	1250	0		
22				899750		

FIGURE 7.19
Final solution for a contractor allocation.

7.7 Coordination of Public Works

Results from long-term planning of civil infrastructure maintenance and rehabilitation (explained in the following chapter) include the optimal schedule of what physical assets to fix and when to fix them on an annual basis. However, raw results from any long-term analysis produce actions randomly scattered across space and time that do not reflect any measures of coordination or operational efficiency, potentially producing many small contracts that would translate into constant disruption of services to the users (public at large) and higher cost to the government (more bids, inspections, relocation of machinery, transporting materials, etc.). Also, uncoordinated actions between different systems (roads and pipes, for instance) may result in utility cuts in the form of premature damage to recently rehabilitated assets (pavements). Therefore, it's in the best interest of departments of transportation and municipalities to prepare medium-range tactical plans that are able to advance or defer interventions across different types of adjacent infrastructure, achieving minimal service disruptions and closure of roads, yet staying close enough to optimal results from the strategic analysis.

In this section, we will learn the basics of space/time coordination applied to civil infrastructure. The coordination of activities is achieved by clustering together some nearby assets that received treatments within a given time window (but not all). When conducting coordination, we need to consider segments within a specific distance (adjacent distance) with interventions scheduled within a given range of time (temporal distance). Temporal distance is the periods of time that an intervention can be deferred or advanced from its original scheduling. These two constraints identify those segments of infrastructure that are candidates and will be possibly cluster together. For example, as illustrated in Figure 7.20, within the prescribed adjacent distance, segments 4 and 10 are receiving treatments 2 and 3, respectively, in year 1, while segment 9 is receiving treatment 3 in year 2, and segment 12 is receiving treatment 1 in year 3. Assuming that the temporal distance is set to

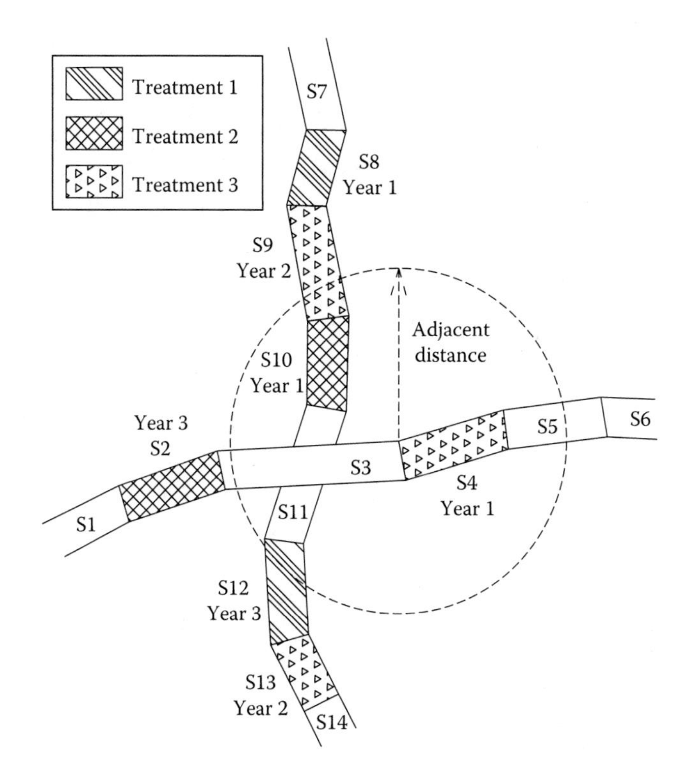

FIGURE 7.20
Coordination of works.

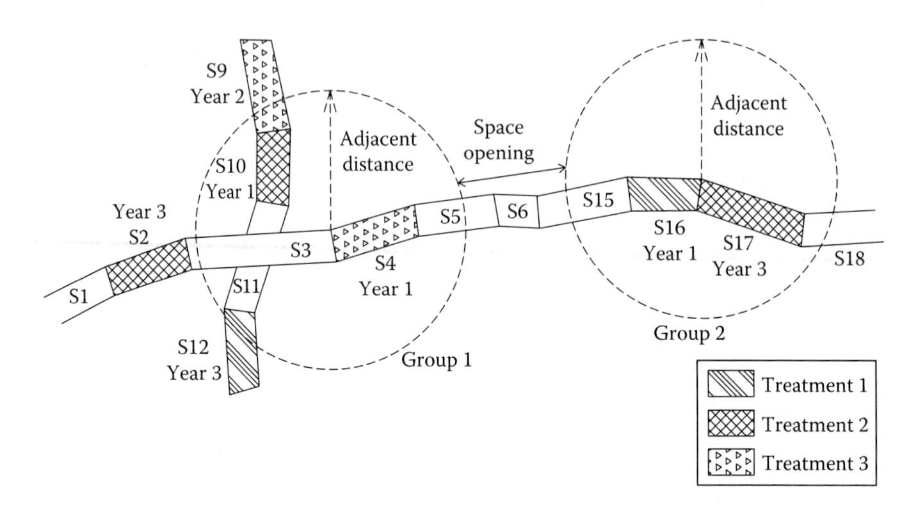

FIGURE 7.21
Coordination: Time and space openings.

2 years, these four segments will be grouped together, creating a new group of asset segments (group 1).

Figure 7.21 illustrates the concepts of time and space openings. Recalling from the previous example, segments 4, 9, 10 and 12 were assigned into group 1; similarly, group 2 could have been formed by joining segments 16 and 17. These two groups can now be joined if they are within a distance called 'space opening', which indicates the willingness of accepting separation between two groups scheduled on the same year if by operational standards it makes more sense to assign them to the same contractor or undertake both projects (groups) at the same time. An extension to this concept is that of 'time opening' in which two groups placed spatially within an acceptable space opening but separated in time (scheduled at different periods) can be joined for similar reasons as noted earlier. This results in a second temporal movement (advance or deferral) of the assets in one of the groups to match the other.

Other elements must be taken into account for performing an analysis capable of developing coordinated tactical plans. Besides spatial and temporal constraints, one must consider the compatibility of actions for the generation of groups (called 'blocks' by the software). Not all maintenance and rehabilitation actions can be implemented together. This consideration depends on the agency's decision, resources, contractor's specialization, compatibility of machinery, time required per task, etc.

7.7.1 Simple Example of Coordination

Let's look at the case of four segments of a road corridor (S1, S2, S3 and S4) receiving various types of maintenance and rehabilitation interventions

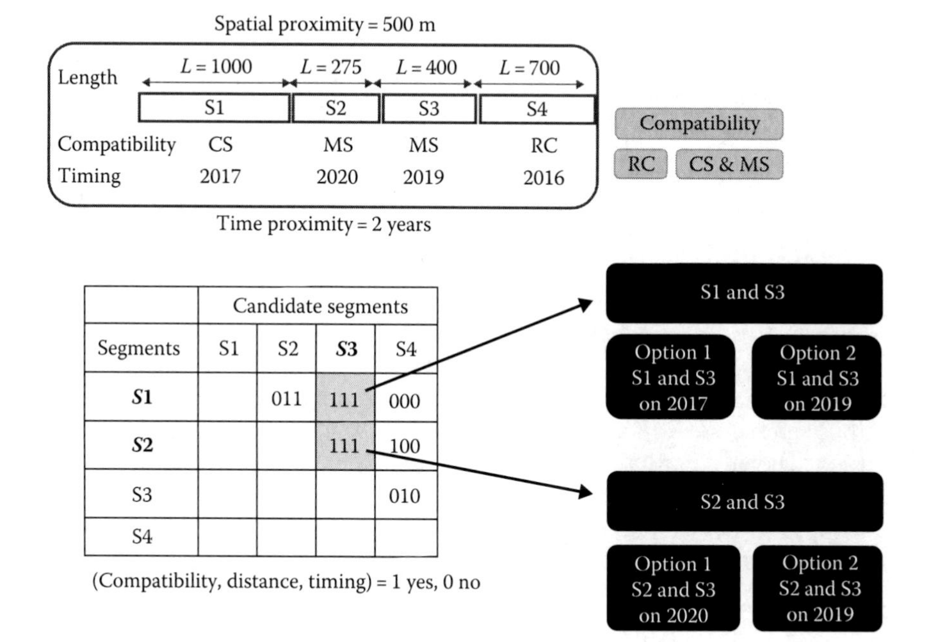

FIGURE 7.22
Coordination set-up.

(Figure 7.22). We will use a binary indicator (0,1) for each criterion, and only pairs of segments with all three criteria (1,1,1) will be chosen as candidates to be scheduled together.

The first step is to apply a filter of compatibility for the interventions. Assume CS and MS are compatible; this initial filter gives a value of zero to segment S4 (receiving RC). Then we look at 500 m adjacency and identify that for segment S1 both S2 and S3 are within such distance. The space criteria will potentially enable us to pack all three segments together. However, we still need to look at time proximity, set to be 2 years in this example. This rules out the possibility of having S1 and S2 together because their timing is separated by 3 years.

The solution to a coordination problem is a set of clustered segments, which satisfy all three criteria (compatibility, distance and timing). When two segments are packed together, you end up with two possible timings for the actual completion of works. Looking at the case of S1 and S3, one can either advance S3 from 2019 to 2017 or defer the timing for S1 from 2017 to 2019, so there are two possible timings. In this case of three segments packed together (not illustrated in the example), there could be up to three options for timing. Selection of the best timing requires another optimization similar to that presented in the following chapter.

7.8 Network Flows

Consider the case of a network of roads and the need for planners to estimate the demand (number of vehicles) at each link of the network. This problem can be partially solved if we count vehicles at each link of the network; however, this is a daunting task. Even if we know the number of vehicles at given points, still sometimes we may be interested in knowing where the trip originated and where it ended. The same is true for different times of the day and the purpose of the trip; however, I won't go into this level of detail.

It is likely that you have some information about the entry and exit points and that you may have some counts at mid-segment points (i.e., segments without entries or exits). Let's call $x_{i,j}$ the flow of traffic entering at i and exiting at j. The summation over all destinations for the flow entering at i must be equal to the entry rate at i. In other words, all vehicles that entered at a specific point ramp to a highway (i) must exit somewhere, so when you add them all up, they should add to that number.

$$\sum_{j=i+1}^{n} x_{i,j} = a_i$$

Figure 7.23 illustrates such an example with concrete numbers of a specific case. The entry rate of 2000 vehicles per hour at node i will exit the system at different nodes: 300 at $i+1$, 600 at $i+2$, etc.

The same is true for all vehicles exiting at a given point (say j); they entered the system somewhere (between 1 and $j-1$).

$$\sum_{i=1}^{j-1} x_{i,j} = b_i$$

Following the previous example, Figure 7.24 illustrates the case of 1000 vehicles per hour (exit rate) at node j that entered the system at different nodes: 500 at $j-3$, 200 at $j-2$ and 300 at $j-1$.

The objective of this type of problem is to minimize the (square of the) differences between the observed and predicted flows. Assume for now you only have one control point located between the nodes i and $i+1$. The summation of all entering flows from 1 to i heading to nodes after $i+1$ minus the observed flow between i and $i+1$ must be minimized:

$$\min \left(\sum_{i=1}^{i} \sum_{i=i+1}^{n} x_{i,j} \right) - q_{i,i+1}$$

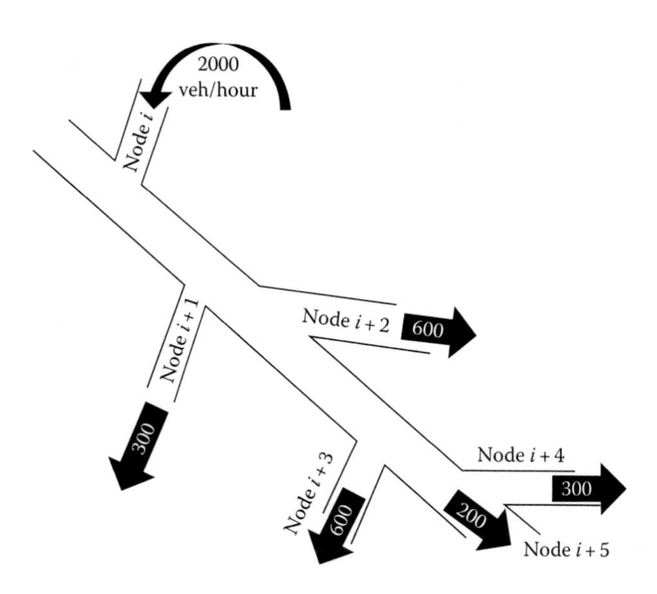

FIGURE 7.23
Example of a vehicle flow.

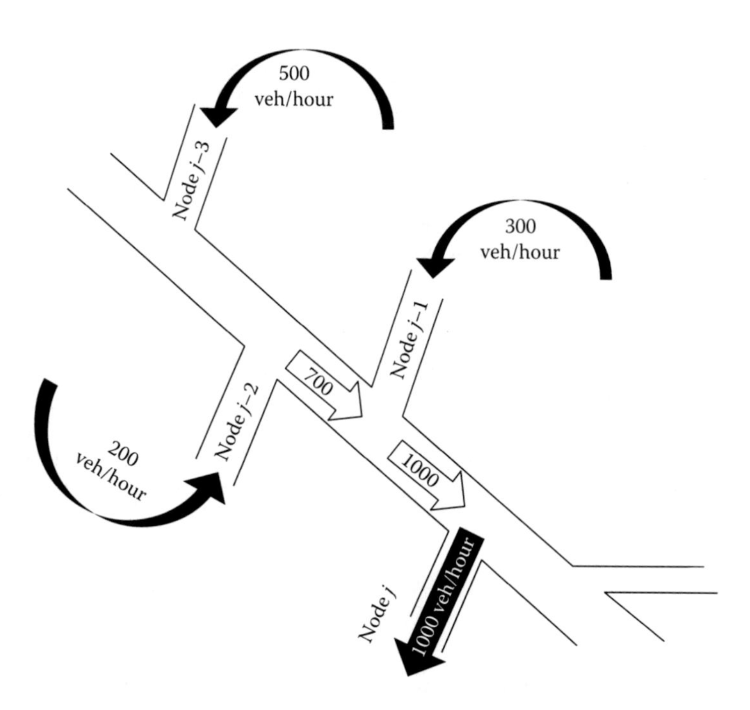

FIGURE 7.24
Vehicle flow in a road network.

In the example presented in Figure 7.24, we have seven nodes; the known flow rate is located between nodes 3 and 4. Hence, the double summation can take on origin the values between 1 and 3 and the destination values between 4 and 7. Let's explicitly state the double summation:

$$x_{1,4} + x_{1,5} + x_{1,6} + x_{1,7} + x_{2,4} + x_{2,5} + x_{2,6} + x_{2,7} + x_{3,4} + x_{3,5} + x_{3,6} + x_{3,7}$$

The observed flow is called $q_{3,4}$, and therefore the objective can be stated as follows:

$$\min(x_{1,4} + x_{1,5} + x_{1,6} + x_{1,7} + x_{2,4} + x_{2,5} + x_{2,6} + x_{2,7} + x_{3,4} + x_{3,5} + x_{3,6} + x_{3,7}) - q_{i,i+1}$$

The desired value of the objective is zero; for this, we can use the Excel solver, and instead of using a minimization, we could set the target value of the objective to be zero.

7.8.1 Example: Pedestrians' Tunnel Network

Let's solve an example of the system of tunnels at Concordia University in Montreal. Figure 7.25 presents a simplified version of it; buildings are represented by circles, and notice that the indexing of nodes is arbitrarily done.

The objective is to minimize the summation of flows between 2, 6, 3, 7 and 1 exiting at 4 or 5, minus the actual flow observed at the segment after 1 and before 4 ($q_{1,4}$) of 1700 pedestrians per hour. However, nodes 2 and 3 are terminal nodes (imagine all students are heading to class at such buildings).

$$\min(x_{6,4} + x_{6,5} + x_{7,4} + x_{7,5} + x_{1,4} + x_{1,5}) - (q_{1,4} = 1700)$$

The objective is that the summation of predicted flows ($x_{6,4} + x_{6,5} + x_{7,4} + x_{7,5} + x_{1,4} + x_{1,5}$) must equal to the observed flow in the segment located between 1 and 4, which is represented at the bottom of the constraints' box. The other constraints can be easily read from Figure 7.26.

Figure 7.27 shows the set-up of solver add-in on Excel. Notice that the objective is not a maximization or a minimization; rather, it is to find a specific value for the summation of predicted flows. In this case, the target value is 1700. The decision variables are selected as the range from *B*2 to *B*14, the type of decision variables is specified to be integers and positive, and then the other constraints are coded as you can see on the interface shown in Figure 7.27.

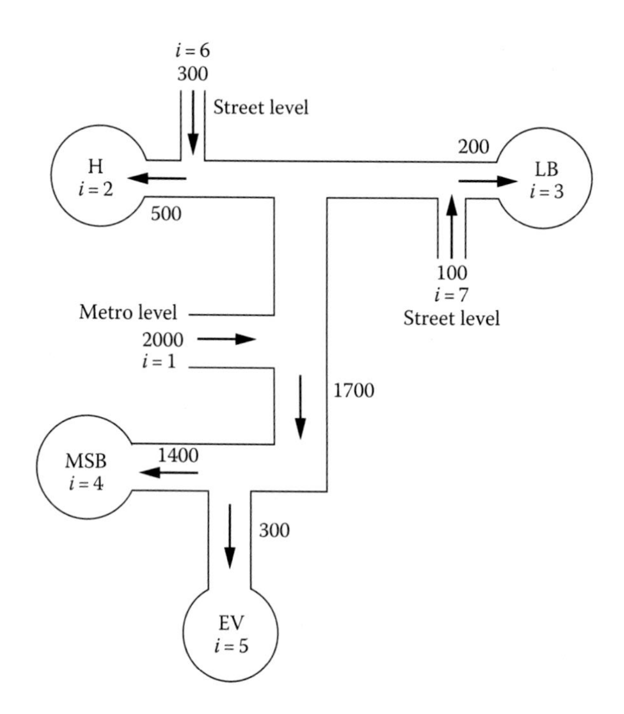

FIGURE 7.25
Pedestrians at underground tunnels.

SUM | =+E9*B2+F9*B3+G9*B4+H9*B5+I9*B6+J9*B7+K9*B8+

	A	B	C	D	E	F	G	H	I	J	K	L	M	N	O	P	Q	R	S
1		X_i,j			x12	x13	x14	x15	x73	x72	x74	x75	x62	x63	x64	x65	value	ineq.	observed
2	x12	500		node 1	1	1	1	1									2000	=	2000
3	x13	200		node 7					1	1	1	1					100	=	100
4	x14	1000		node 6									1	1	1	1	300	=	300
5	x15	300		node 2	1					1			1				500	=	500
6	x73	0		node 3		1			1					1			200	=	200
7	x72	0		node 4			1				1				1		1400	=	1400
8	x74	100		node 5				1				1				1	300	=	300
9	x75	0		Goal=1700			1	1			1	1			1	1	=+E9*	=	1700
10	x62	0					Constraints												
11	x63	0																	
12	x64	300																	
13	x65	0																	

FIGURE 7.26
Excel set-up for underground tunnels.

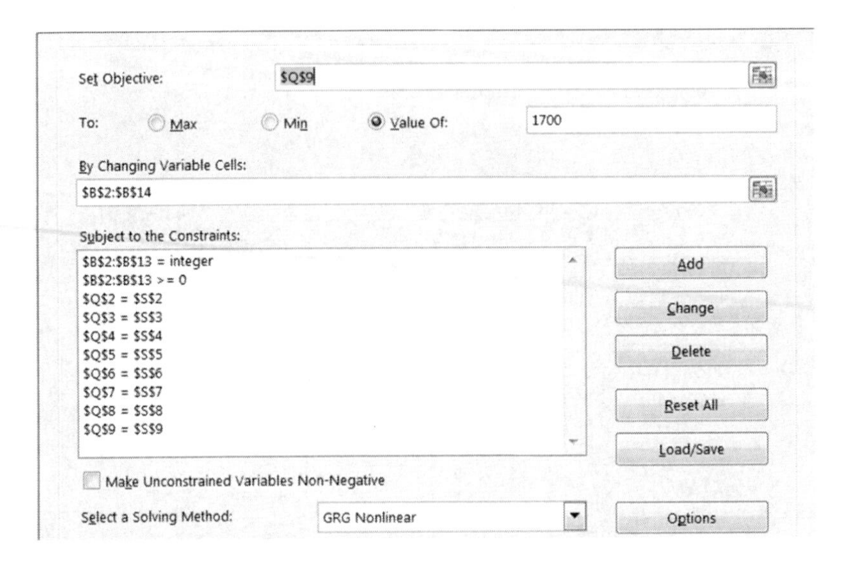

FIGURE 7.27
Solver set-up for underground tunnels.

Exercises

1. A traveller is trying to go from node 4 to node 1 (Figure 7.28). Formulate and solve the shortest path problem using travel time as cost.
2. Prisoners are planning to escape from jail in a dictatorial country. Help them plan the escape route: identify number of prisoners per link and the overall route to follow (use a trans-shipment formulation). Figure 7.29 shows the network of escape tunnels (one direction of movement). Travellers start at nodes 4 (12 travellers), node 5 (6 travellers) and node 6 (13 travellers). All travellers are going to node 1 (liberty!). Travel time for every link along with a sense of strictly permitted movement is shown over every link. Formulate the problem.
3. Modify the formulation of Exercise 7.1 and reverse the direction of arrows; the departure point now is node 1 and the arrival point is node 4.
4. Modify the formulation of the objective by removing the limitation that nodes 2 and 3 are terminal nodes in Example 7.8.1.
5. Formulate and solve the network in Figure 7.30.
   The solution is shown in Figure 7.31.

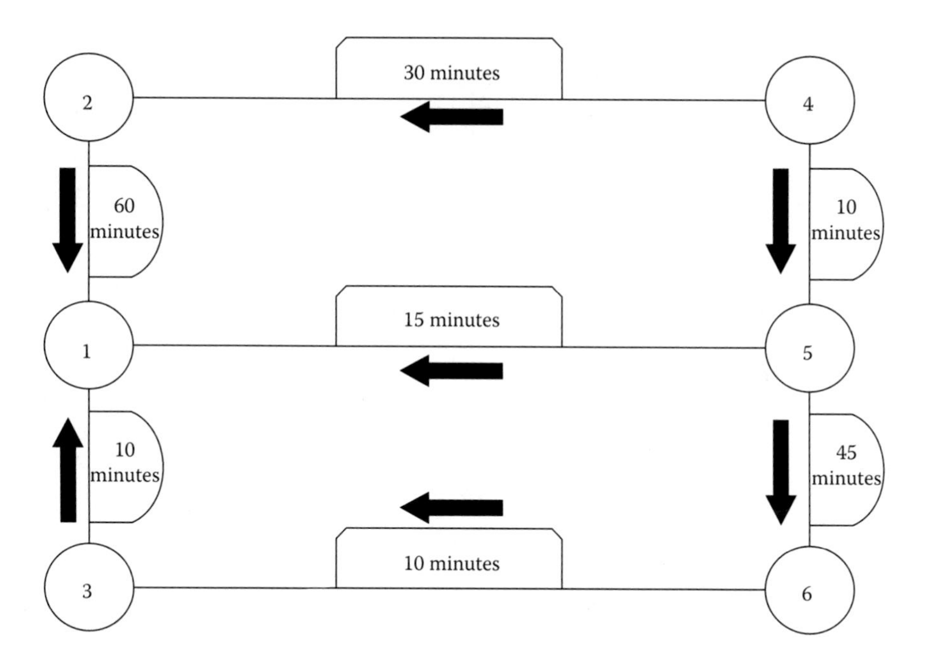

FIGURE 7.28
Exercise 7.1.

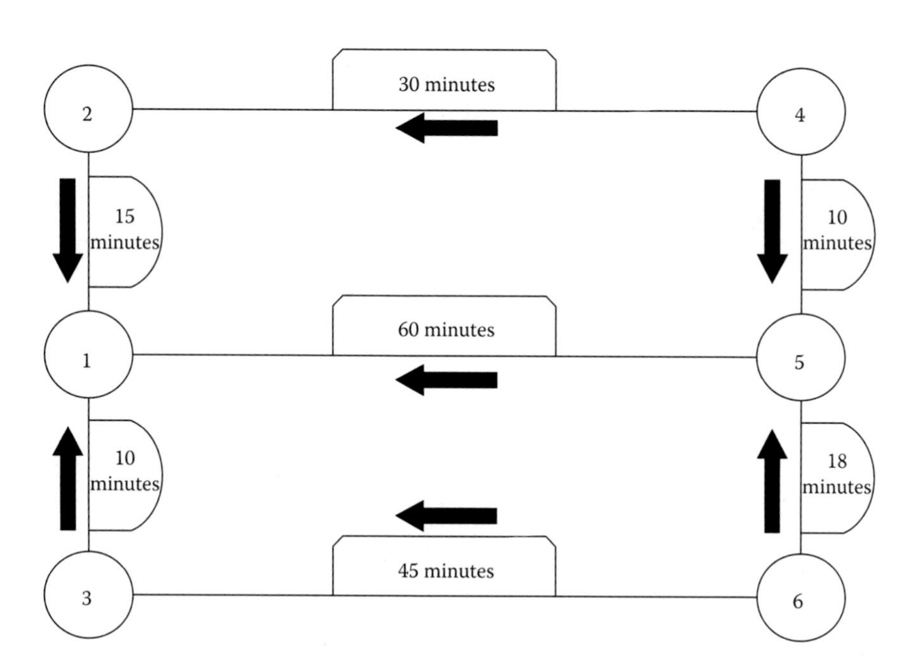

FIGURE 7.29
Exercise 7.2.

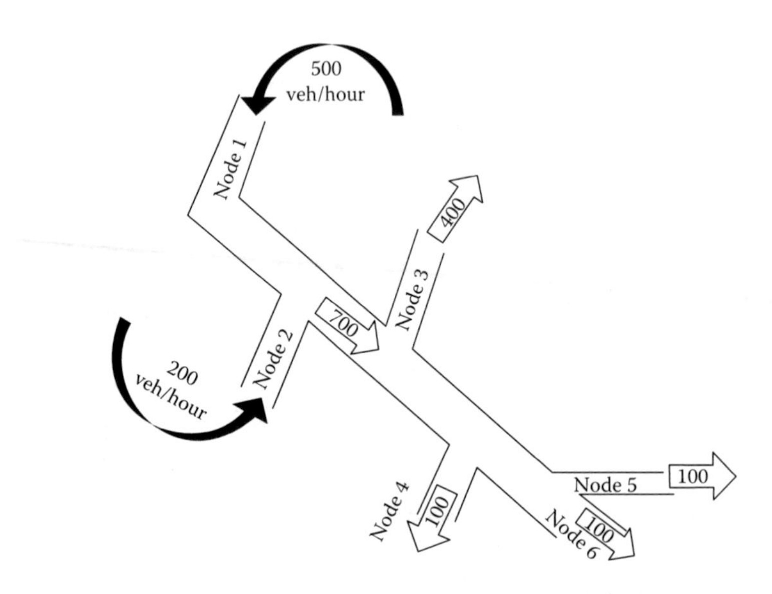

FIGURE 7.30
Exercise 7.5.

Solutions

1. Objective MIN cost $= \text{MIN } x_{42}c_{42} + x_{45}c_{45} + x_{21}c_{21} + x_{51}c_{51} + x_{56}c_{56} + x_{63}c_{63} + x_{31}c_{31}$

   Constraints

   Node 1 : $x_{21} + x_{51} + x_{31} = 1$ (because it is only one traveler)

   Node 2 : $x_{42} = x_{21}$

   Node 3 : $x_{63} = x_{31}$

   Node 4 : $x_{42} + x_{45} = 1$ (one traveler)

   Node 5 : $x_{45} = x_{51} + x_{56}$

   Node 6 : $x_{56} = x_{63}$

   $X_{ij} \geq 0$

2. Objective MIN cost $= \text{MIN } x_{42}c_{42} + x_{45}c_{45} + x_{21}c_{21} + x_{51}c_{51} + x_{65}c_{65} + x_{63}c_{63} + x_{31}c_{31}$

   Constraints

   Node 1 : $x_{21} + x_{51} + x_{31} = 31$ (because there are 31 in total)

   Node 2 : $x_{42} = x_{21}$

   Node 3 : $x_{63} = x_{31}$

   Node 4 : $x_{42} + x_{45} = 12$

   Node 5 : $x_{45} + x_{65} + 6 = x_{51}$

Node 6 : $x_{63} + x_{65} = 13$

$X_{ij} \geq 0$

3. Objective MIN cost $=$ MIN $x_{24}c_{24} + x_{54}c_{54} + x_{12}c_{12} + x_{15}c_{15} + x_{65}c_{65} + x_{36}c_{36} + x_{13}c_{13}$

   Constraints

   Node 1 : $x_{12} + x_{15} + x_{13} = 1$ (because it is only one traveler)

   Node 2 : $x_{24} = x_{12}$

   Node 3 : $x_{36} = x_{13}$

   Node 4 : $x_{24} + x_{54} = 1$ (one traveler)

   Node 5 : $x_{54} = x_{15} + x_{65}$

   Node 6 : $x_{65} = x_{36}$

   $X_{ij} \geq 0$

4.

$$\min(x_{2,4} + x_{2,5} + x_{6,4} + x_{6,5} + x_{3,4} + x_{3,5} + x_{7,4} + x_{7,5} + x_{1,4} + x_{1,5})$$
$$- (q_{1,4} = 1700)$$

5.

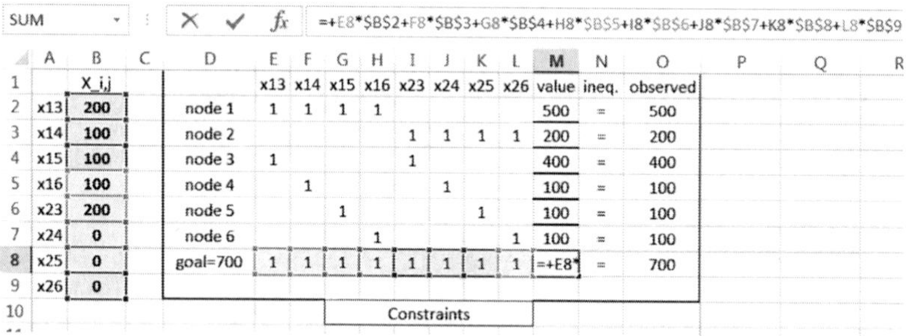

SUM | × ✓ *fx* | =+E8*B2+F8*B3+G8*B4+H8*B5+I8*B6+J8*B7+K8*B8+L8*B9

	A	B	C	D	E	F	G	H	I	J	K	L	M	N	O	P	Q	R
1		X_i,j			x13	x14	x15	x16	x23	x24	x25	x26	value	ineq.	observed			
2	x13	200		node 1	1	1	1	1					500	=	500			
3	x14	100		node 2					1	1	1	1	200	=	200			
4	x15	100		node 3	1				1				400	=	400			
5	x16	100		node 4		1				1			100	=	100			
6	x23	200		node 5			1				1		100	=	100			
7	x24	0		node 6				1				1	100	=	100			
8	x25	0		goal=700	1	1	1	1	1	1	1	1	=+E8*	=	700			
9	x26	0																
10								Constraints										

FIGURE 7.31
Solution to Exercise 7.5.

8

Civil Infrastructure Management

8.1 Introduction

This chapter introduces the reader to a key contemporary problem: the allocation of maintenance and rehabilitation (MR) to sustain a network of infrastructure assets in good condition. The ideas of this chapter can be extended to account for environmental pollution, user's safety, cost and travel time, among others. A brief indication of how to achieve this is provided at the end of this chapter. However, to simplify the concepts, I choose to concentrate the explanation for roads, specifically pavements. The rest of this book concentrates on explaining how to structure the problem and illustrates one possible way of solving it.

8.2 Dynamic Effect of Decisions on Time

So far, in this book, decisions have had a singular immediate effect; however, by nature the choices related to civil infrastructure management are time dependent (formally called 'dynamic'). The condition of any physical infrastructure decays across time; the longer you wait to apply an intervention, the more the deterioration of the asset and the more it is going to cost you to repair it. Then, why do governments wait until last minute to fix them? The answer lies on the overwhelming size of the assets and the daunting task of using a small budget to satisfy so many repair needs. Let's explore together the application of some of the techniques learned so far to solve this problem: First, we use a linear regression to construct the deterioration performance of a given pavement type. Then we formulate the problem and apply optimization to solve it. Note you can apply linear programming (dynamic-mixed integer) or a heuristic approach as we propose here.

Think of a relatively new road with some cracks; the cracks will allow water into the base, which will produce erosion of the foundation and lead to potholes and settlements. That is, of course, if you do not repair it before the rainy season. Let's assign a value on a 0–100 scale to the condition of the road, say 90. Next, year you will start to observe some deformation on the surface, so say condition drops to 80; if you do nothing (no intervention) the condition will continue to drop to say 70. At this point the lack of any intervention will have major problems. The fine-graded aggregates of the base will be gone, and the pavement will start to break in blocks and pop up, so condition will now drop by 20 instead of 10, to a value of 50, then 30, then 0. Table 8.1 illustrates the annual condition progression for this road segment. As you can imagine, at the beginning the only needed treatment is to rout and seal the cracks (we call this 'crack sealing' and denote it as CS). Then in year 2, you start to have deformation that requires a thin overlay called 'micro-surfacing' (MS); after year 3, you have a more significant deformation that requires an overlay, after year 4, the base is no longer as strong as should be, and you have to remove the asphalt cement (AC) layer and re-compact the base. After year 5, you need to redo the road, that is, a full reconstruction must be performed. The real rate of deterioration of a road depends on the number and weight of trucks, the amount of rain and freeze–thaw cycles, the type of soil (swelling) and the quality of the materials used for the pavement structure. We will assume that we have isolated a group of roads, all of which are exposed to very similar levels of traffic and all are in the same environmental region, with almost the same type of soil and materials. Such a group is called 'homogeneous' given that they have similar characteristics in nature.

The deterioration behaviour of this homogeneous group along with associated cost of interventions is shown in Figure 8.1. As can be seen, the cost intervals define the operational windows for the application of treatments

TABLE 8.1

Sample Inventory of Roads

Year/Asset	B	C	D	E	F	G	H
1	90	91	88	89	90	91	89
2	80	82	79	80	81	79	80
3	69	70	71	70	69	70	70
4	60	61	60	59	60	59	60
5	50	51	50	49	50	49	50
6	40	41	40	39	40	39	40
7	30	31	30	29	30	29	30
8	20	21	20	19	20	19	20
9	10	11	10	9	10	9	10
10	0	1	0	0	0	0	0

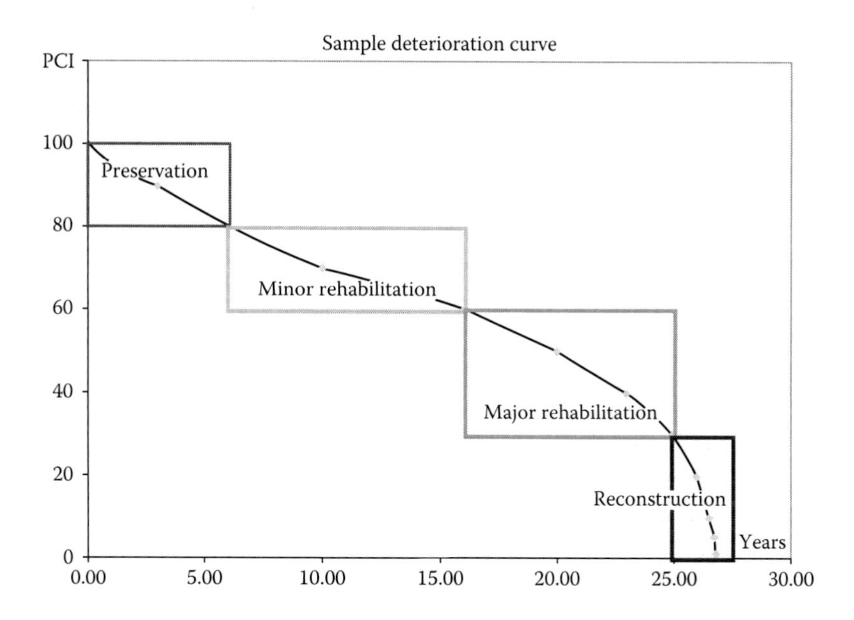

FIGURE 8.1
Roads' deterioration and cost.

and interventions; the longer you wait, the more expensive it becomes, and even if you fix the road today, in a few years the road will have to be fixed again. So the times elapsed from today and those between interventions are also important. In a nutshell, time is important because it represents an exponential shift in cost if you wait too long, and a small fraction of your budget if you keep the good roads in good condition.

8.3 Performance Modelling

Performance refers to the behaviour of an asset, whether this indicates improvement or decay. Performance is typically based upon the very own characteristics of the asset, and it is affected by the intensity of the loads applied to it, the strength of its structure (which relates to the materials and quality of the construction method) and the aggressiveness of the environment. Performance can be measured through a deterministic or probabilistic approach. A deterministic method is simply the one that represents the mean or the average, but a probabilistic method assigns probabilities to future values. We will study sample methods for deterioration and improvement. How to integrate performance into the problem will be explained later in this chapter.

8.3.1 Deterioration Modelling

This section illustrates two methods to construct a deterioration curve. A deterioration curve is one that serves to express the dynamic nature of the problem at hand and hence its relevance. For simplicity of the example, let's assume that we are still dealing with pavements and that road's characteristics remain unchanged. Let's start with a familiar method: linear regression. Given that we have several segments on our (homogeneous) group, we need to capture their performance with a curve. Table 8.1 presents the observed condition values for a fictional group of pavements used in this example. The time scale has been broken and pavement condition matched such that all segments start at about the same level. Not all pavements remained untreated for the whole 14-year period of Table 8.1; this is why you see variable length of information on the columns of the table.

Generally speaking, the application of linear regression results in an equation that could take a linear, exponential, logarithmic, power or polynomial form, whichever fits best the observed data. For the given information, we could apply the methods explained in Chapters 4 and 5. For rapid calculation, you can make use of Excel to obtain the best fit, which is given by the command 'Add trend line', and then select the display equation and R^2 value. For the given example, you will find that the equation is $y = 100 - 10x$ after the rounding terms (Figure 8.2).

We can also apply a transition probability matrix to capture the movements of deterioration. The matrix will have the form of a discrete mechanism, which will connect the deterioration values with service life of the asset. Movements away to the right from the main diagonal represent deterioration of the entire years on the lifespan. As the asset gets older, the deterioration

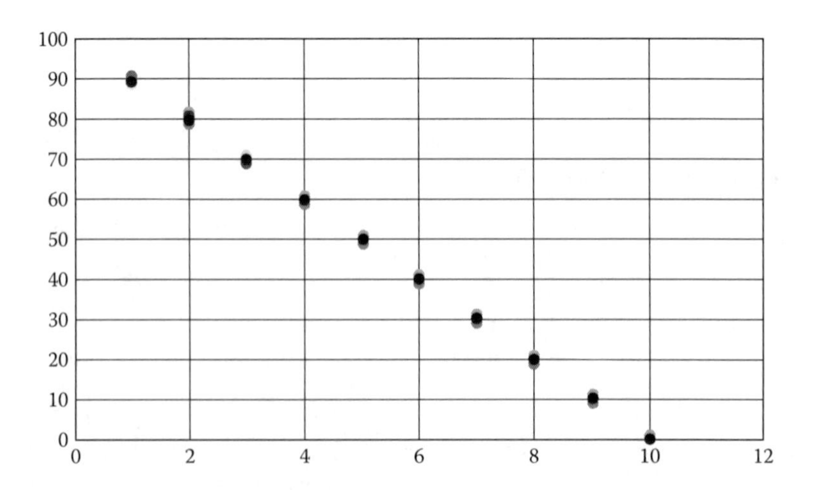

FIGURE 8.2
Linear deterioration.

TABLE 8.2

Deterioration

Age	1	2	3	4	5	6	7	8	9	10	11	12	13	14
1	90	10	0	0	0	0	0	0	0	0	0	0	0	0
2	0	80	20	0	0	0	0	0	0	0	0	0	0	0
3	0	0	50	40	10	0	0	0	0	0	0	0	0	0
4	0	0	0	20	60	20	0	0	0	0	0	0	0	0
5	0	0	0	0	0	90	10	10	0	0	0	0	0	0
6	0	0	0	0	0	0	85	10	5	0	0	0	0	0
7	0	0	0	0	0	0	0	80	10	10	0	0	0	0
8	0	0	0	0	0	0	0	0	60	20	10	10	0	0
9	0	0	0	0	0	0	0	0	0	40	40	10	10	0
10	0	0	0	0	0	0	0	0	0	0	10	80	5	5

rate increases. For instance, Table 8.2 presents a sample 10-year service life matrix in which brand new assets at age 1 tend to remain in the same condition interval, with only 10% deterioration in 1 year. During year 2, 80% of the assets remains unchanged, and 20% deteriorates in 1 year. Half of those at age 3 deteriorated, with most (40%) dropping 1 year in condition and 10% dropping 2 years. As we move down, the matrix assets tend to deteriorate more and more as well as the spread tends to increase. Notice that the sum of each row must add to 100%.

8.3.2 Improvement Modelling

The modelling of improvement can be based on experience, expert criteria or a measurement of an asset characteristic before and after the study. Table 8.3 illustrates the possible values for improvement for roads. For the purpose of this book, the rate of deterioration after the application of an intervention will remain the same.

There are two other ways to estimate the effectiveness of a treatment: by measuring the jump in condition by looking into a specific indicator such as cracked area, rut depth and roughness. Presume you have measurements of

TABLE 8.3

Intervention Effectiveness (additional life span)

Intervention	Effectiveness (Years)	Rate
Crack sealing	1	Same
MS	6	Same
Overlay	10	Same
Reconstruction	20	Same

TABLE 8.4

Intervention Effectiveness (indicator increase)

Intervention	Effectiveness	Rate
Crack sealing	$IRI_t - 0.2$	Same
MS	$IRI_t - 1.1$	Same
Overlay	Reset to 1.2	Same
Reconstruction	Reset to 0.8	Same

the International Roughness Index (IRI) for a dozen of segments; the lower the value of roughness, the smoother the surface, and hence improvements can be seen when the value of roughness drops. Assume that we can identify each type of treatment. Then the only thing remaining is to average the improvement between the values before and the after intervention to obtain the effectiveness. Table 8.4 is similar to Table 8.3, but this one is given in terms of the condition indicator.

Some interventions are intended to smooth the surface, whereas others are to fix the structure, and reconstruction is for both purposes. Crack sealing will not change much the smoothness but will slow down the rate of deterioration of the structure (just as painting a steel bridge). MS will smooth the surface to some extent but not to a brand new condition. Reconstruction will definitively return the surface to a new condition state. Measuring the variability on the before versus after condition for those treatments that return the surface to a new-like condition will result in highly variable numbers, and the average will not be a good indicator. Rather, we reset the value of roughness to a given level.

For those cases where no record of applied treatment exists, one can use a method based on an inverted Markov chain to identify operational windows and capture the effectiveness for each intervention type. The method departs from the same principle, observes and registers the before versus after movements but uses a very specific grid to register such numbers. The grid is tuned to match discrete movements on the apparent age of the given asset. The concept of apparent age comes from the deterioration curve, which connects any condition indicator with time movements in years. The grid used to identify operational windows and measure treatment effectiveness corresponds to a simple table tuned to intervals of condition that correspond to 1-year movements of the condition.

An example will illustrate this concept. Assume we have separated from the database all segments that exhibit improvement. Suppose also that the deterioration curve can be represented by $PCI_t = 100 - 10t$, that is, at time 0 the pavement condition index (PCI) value is 100, and at time 10 the PCI value is 0. If we wish to build a 10×10 table, we can do that by having 10 intervals; that is, 0–10, 10–20, . . . , until 90–100. These intervals correspond to the ages of 10, 9, . . . , 1 year. Table 8.5 illustrates such case. The presented

TABLE 8.5
Measuring Improvement

Age	1	2	3	4	5	6	7	8	9	10
1	100	0	0	0	0	0	0	0	0	0
2	80	20	0	0	0	0	0	0	0	0
3	0	100	0	0	0	0	0	0	0	0
4	0	90	10	0	0	0	0	0	0	0
5	0	0	90	10	0	0	0	0	0	0
6	80	15	5	0	0	0	0	0	0	0
7	0	100	0	0	0	0	0	0	0	0
8	100	0	0	0	0	0	0	0	0	0
9	100	0	0	0	0	0	0	0	0	0
10	100	0	0	0	0	0	0	0	0	0

values are just for illustrative purposes and should not be used on real-life applications; rather, the method proposed here is used to estimate them.

Counting the number of jumps from the main diagonal to the left up until the cells with values different from zero is equivalent to counting the number of years (and associated likelihood) that the treatment will extend the service life of the asset. At age 1 (new), nothing changes, at age 2, 20% remains unchanged and 80% gains 1 year of life, at age 3 everybody extends their life by 1 year, and at age 4 (and 5) the jump increases to 2 years for 90% of the observations. This change marks the possibility of dealing with a different type of intervention (MS perhaps). At age 6, the jump increases again, this time to 5 years additional life to 80% of the observations; the same gain is observed for those at age 7. Finally, at ages 8, 9 and 10, we observed gains of 7, 8 and 9 additional years. In other words, reconstruction brings everybody at that range back to a new-like condition. The newly learned information can be compacted on a table with generic treatments and operational windows as shown before.

8.4 Infrastructure Management

8.4.1 Preamble: Joining Deterioration and Improvement

This section deals with the formal definition of the problem. We will start by looking into one segment or asset alone for two periods and then expand it to many segments and multiple periods. The connection between two contiguous periods is central to the structure of the problem. The condition of an asset today depends on its condition yesterday, and if no intervention has been applied, then the condition will be the same or worse. When a whole year has elapsed, the condition will surely worsen. If we call Q_t the condition at time t, then the condition at time $t + 1$ (Q_{t+1}) is equal to the condition of the year before minus the deterioration rate from 1 year to another (D_t). If an intervention has been applied, then the condition could jump up and actually increase. Equations 8.1 and 8.2 show the two basic but very important characteristics of the condition's time dependency.

$$Q_{t+1} = Q_t - D_t \text{ condition next year if asset deteriorates} \tag{8.1}$$

$$Q_{t+1} = Q_t + I_t \text{ condition next year if asset improves} \tag{8.2}$$

Dealing with multiple assets requires us to have a way to handle both possibilities as one. Let's call x_t the decision variable of whether a treatment is applied or not; then $1 - x_t$ will be the contrary, that is, when a treatment is not applied, it returns a value of 1. Equation 8.3 combines both Equations 8.1 and 8.2 into one.

$$Q_{t+1} = (1 - x_t)(Q_t - D_t) + x_t(Q_t + I_t) \tag{8.3}$$

This creates a mechanism to link all periods in an interdependent manner. This mechanism is at the core of any decision-making system that attempts to optimize condition. It utilizes both deterioration and improvement modelling. As can be seen, every year an asset can undergo deterioration (if no intervention is applied) or improvement (on the contrary). Consider the case of two assets in three periods (current and two in the future); Figure 8.3 illustrates such case.

From Figure 8.3, we can see that possible strategies for the MR are based on combinations of courses of action across time and across assets. As seen in the figure, one uses such a dynamic mechanism to predict all courses of action and then one works his or her way by restricting the courses of action that are feasible from a budget perspective. Hence, there is a need to obtain two aggregated indicators: one for the total cost and another one for the total condition (network-wide).

As seen in Figure 8.3, the number of possible combinations increases following a power function given by the number of treatments to the power of time multiplied by the number of assets. Dealing with many assets and many periods of time is impossible for hand calculations; only computers have the power to do so. In Section 8.5 at the end of this chapter, I provide you with the codes and structure to formulate this problem in Excel; however, the solution is a little more difficult but possible.

Cost follows a similar scheme. If an intervention is scheduled, then the cost is equal to the unitary cost of such intervention times the length, area or size of the asset. For instance, replacing a Jersey barrier would cost \$250,000/km, overlaying a pavement (full mill of AC surface) would cost around \$28/m^2. Cost could be used as a constraint or as an objective; this is in line with what

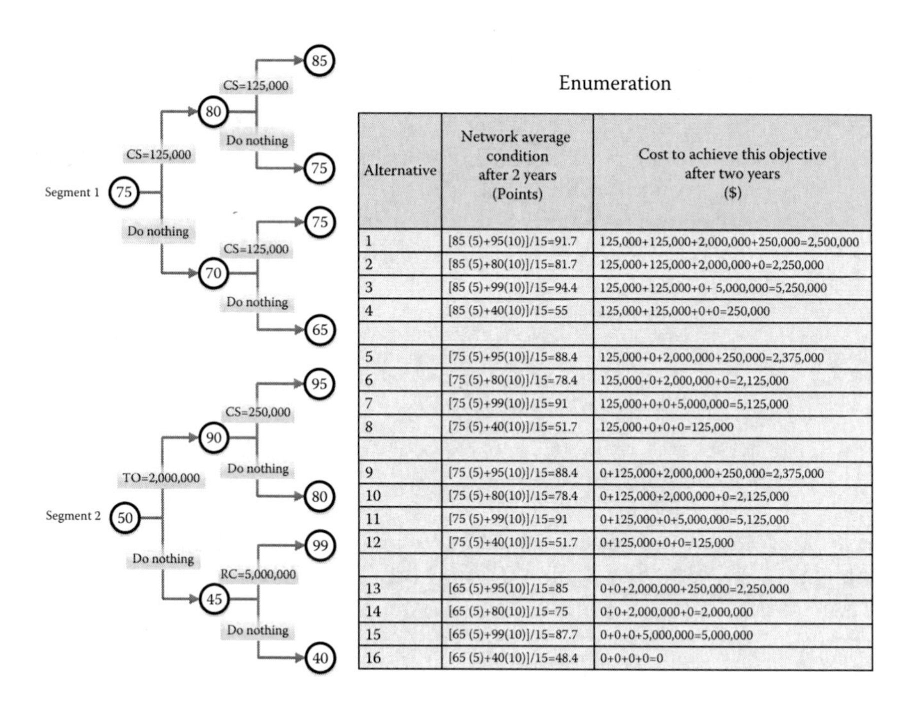

Alternative	Network average condition after 2 years (Points)	Cost to achieve this objective after two years ($)
1	[85 (5)+95(10)]/15=91.7	125,000+125,000+2,000,000+250,000=2,500,000
2	[85 (5)+80(10)]/15=81.7	125,000+125,000+2,000,000+0=2,250,000
3	[85 (5)+99(10)]/15=94.4	125,000+125,000+0+ 5,000,000=5,250,000
4	[85 (5)+40(10)]/15=55	125,000+125,000+0+0=250,000
5	[75 (5)+95(10)]/15=88.4	125,000+0+2,000,000+250,000=2,375,000
6	[75 (5)+80(10)]/15=78.4	125,000+0+2,000,000+0=2,125,000
7	[75 (5)+99(10)]/15=91	125,000+0+0+5,000,000=5,125,000
8	[75 (5)+40(10)]/15=51.7	125,000+0+0+0=125,000
9	[75 (5)+95(10)]/15=88.4	0+125,000+2,000,000+250,000=2,375,000
10	[75 (5)+80(10)]/15=78.4	0+125,000+2,000,000+0=2,125,000
11	[75 (5)+99(10)]/15=91	0+125,000+0+5,000,000=5,125,000
12	[75 (5)+40(10)]/15=51.7	0+125,000+0+0=125,000
13	[65 (5)+95(10)]/15=85	0+0+2,000,000+250,000=2,250,000
14	[65 (5)+80(10)]/15=75	0+0+2,000,000+0=2,000,000
15	[65 (5)+99(10)]/15=87.7	0+0+0+5,000,000=5,000,000
16	[65 (5)+40(10)]/15=48.4	0+0+0+0=0

FIGURE 8.3
Total enumeration example.

you learned before in previous chapters: any constraint (objective) can be re-expressed as an objective (constraint). Let's have a look at one of the many possible ways to formulate this optimization problem.

8.4.2 Optimization Algorithms

An optimization algorithm is nothing but a mathematical expression for the objective and the constraints. For the problem at hand (i.e. optimizing condition and cost), there are several possible ways to define it. Let's start by defining the problem when the level of budget is known and therefore we are constrained by it. In this case, the decision maker wishes to maximize the level of condition of its assets with the given level of funding per year. The objective (Equation WW) is to maximize the total network condition, which is given as a summation across all assets, and this objective can be repeated across time as well, given the impact of the decisions across time, as explained before.

$$\max_{x_{t,i}} \sum_{t=1}^{T} \sum_{i=1}^{N} L_i Q_{t,i}$$

As you can notice, there is no decision variable explicitly expressed in the aforementioned expression. Now, remember that we have seen an explicit expression for condition that carries both improvement and deterioration along with the decision variable. Let's include it in the algorithm. Note that such an expression will be multiplied by the length of each segment as we wish to obtain an indication of the total level of condition for the entire network and the contribution of a segment to the total is proportional to its length.

$$\max_{x_{t,i}} \sum_{t=1}^{T} \sum_{i=1}^{N} L_i[(1 - x_{t,i})(Q_{t,i} - D_{t,i}) + x_{t,i}(Q_{t,i} + I_{t,i})]$$

The main constraint for this problem is the total cost resulting from the amount of money spent on each asset (hence the summation across them). If the budget level is fixed, then it could be expanded to include all periods of time in the analysis.

$$\sum_{t=1}^{T} \sum_{i=1}^{N} L_i c_{t,i} x_{t,i} \leq B_t$$

When the budget changes every year, the expression for its constraint can only contain a summation across assets, and multiple expressions (one per period of time) would be specified. Assume that the budget values are given by B_1, B_2, B_3, B_4 *and* B_5 and that after year 6 the level of budget remains constant at a value of $B_{t \geq 6}$. The following equation illustrates such a case:

$$\sum_{i=1}^{N} L_i c_{t,i} x_{t,i} \leq B_{t=1}$$

$$\sum_{i=1}^{N} L_i c_{t,i} x_{t,i} \leq B_{t=2}$$

$$\sum_{i=1}^{N} L_i c_{t,i} x_{t,i} \leq B_{t=3}$$

$$\sum_{i=1}^{N} L_i c_{t,i} x_{t,i} \leq B_{t=4}$$

$$\sum_{i=1}^{N} L_i c_{t,i} x_{t,i} \leq B_{t=5}$$

$$\sum_{t=6}^{T}\sum_{i=1}^{N} L_i c_{t,i} x_{t,i} \leq B_t$$

A possible variation to this algorithm is given by the need to satisfy specific requirements, sometimes coming from political campaign promises, or from direct orders from the government. Suppose a 5% improvement is required for a period of 5 years.

Alternatively, we could be facing another famous specification that requires a group of assets to be reduced every year (those in poor condition). This requirement brings the need to utilize an indexing system to filter the target group. The condition of the network at specific moments of time needs to be equivalent to a specific level.

8.5 Solving Optimization

Two general approaches can be used to solve this type of optimization: a linear programming (dynamic binary) approach or a heuristic method. The size of the problem (number of possible courses of action) determines the selection of the method.

First, we need to take advantage of the indexing system to aggregate across groups. This is equivalent to having one asset type per group, but this asset will contain many individual assets (or segments) together, which facilitates the computational task. The larger the number of filters, the more likely the optimal solution found matches that of a model based on individual handling of assets, but if the indexing returns very few groups, then we could run into a suboptimal solution that is not even capable of spending the total available budget.

The aggregation of individuals is commonly done in economics when handling recursive problems; in a similar manner, we will extensively use aggregation in order to reduce the computational burden.

One needs to bear in mind that when dealing with hundreds (if not thousands) of segments, and tens of years, there is a need to use certain strategies to ease the computational task. The first trick, as explained previously, is the aggregation of similar segments using the filters to obtain one representative individual per group. The second trick, also previously learned, is to use the constraints to reduce the number of possible courses of action by removing those that are infeasible. The solution will be based on identifying the most cost-effective set of actions. This could be done each year, and hence the

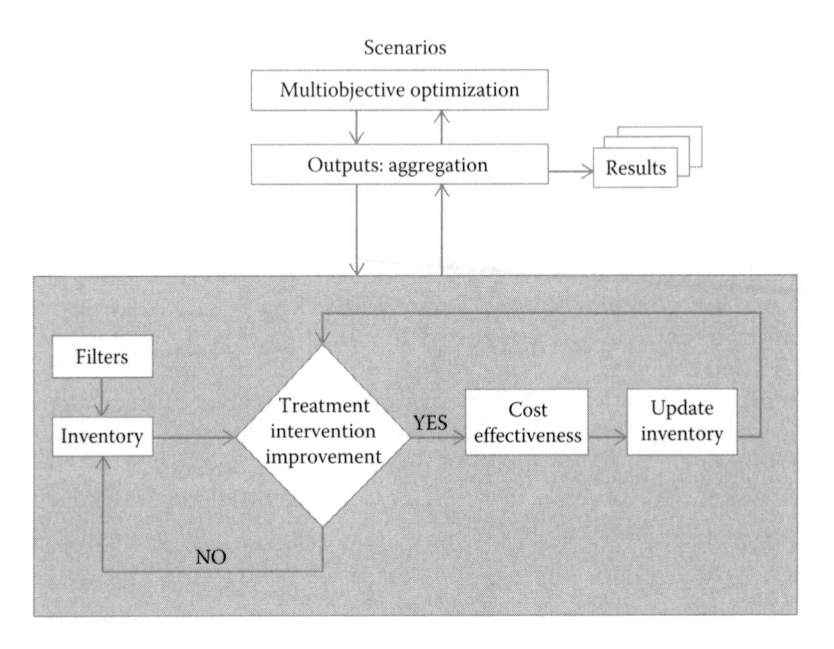

FIGURE 8.4
Overall method.

inventory will be updated every year. The outputs aggregated and results generated are shown in Figure 8.4. Let's see how that actually happen.

Let's assume that we have already aggregated the individual assets and we are currently dealing with representative individuals. As can be seen from Equation 8.3, the decision variable x_t needs to be found for each possible combination (Figure 8.5). One way to approach the solution is by simply testing binary values for the decision variable until the value of the objective cannot be improved further. A heuristic algorithm, such as a genetic algorithm or evolutionary algorithm, can be used for this purpose. In addition, we need a stopping criterion. Other algorithms such as simulated annealing would have the stopping criterion incorporated as a mechanism within the approach to the solution.

Excel is only able to provide a representation of the problem but is unable to solve it. For finding a solution, we need to connect Excel with an external solver, such as Cplex, Lindo, Mosek and AMPL, or, in our case, a solver risk platform, which is a low-cost engine capable of solving an annual optimization problem. Hence, we need to solve year by year in sequence in order to obtain the final solution. An alternative platform (such as AMPL or MATLAB®) for representing the problem could be used, but we choose Excel for convenience given its popularity and ease of use.

Figure 8.6 illustrates the extended set-up used in Excel that connects several periods (the figure only shows three but the model actually works

AR	AS	AT	AU	AW	AX	AY	AZ	BB	BC	BD	BE
x	OWTreat 2018	PCI 2018	Cost 2018	x	OWTreat 2019	PCI 2019	Cost 2019	x	OWTreat 2020	PCI 2020	Cost 2020
0	Do nothing	97.944	0	1	Preserve	100	1000	1	Preserve	100	1000
1	Preserve	100	500	1	Preserve	100	500	1	Preserve	100	500
1	Preserve	100	51170	0	Do nothing	99.182549	0	0	Do nothing	97.944	0
0	Do nothing	99.1825	0	1	Preserve	100	500	1	Preserve	100	500
1	Preserve	93.9164	12500	1	Preserve	98.916403	12500	0	Do nothing	96.0675	0
1	Preserve	100	13615	0	Do nothing	99.182549	0	1	Preserve	100	13615
0	Do nothing	99.1825	0	0	Do nothing	97.944006	0	1	Preserve	100	24500
1	Preserve	100	38500	1	Preserve	100	38500	0	Do nothing	99.1825	0
1	Preserve	100	2500	1	Preserve	100	2500	1	Preserve	100	2500
1	Preserve	100	17585	0	Do nothing	99.182549	0	1	Preserve	100	17585
0	Do nothing	97.944	0	1	Preserve	100	2000	1	Preserve	100	2000
0	Do nothing	99.1825	0	1	Preserve	100	14500	1	Preserve	100	14500
1	Preserve	100	10000	1	Preserve	100	10000	1	Preserve	100	10000
0	Do nothing	97.944	0	1	Preserve	100	9000	1	Preserve	100	9000
1	Preserve	100	54000	1	Preserve	100	54000	0	Do nothing	99.1825	0
1	Preserve	100	500	0	Do nothing	99.182549	0	0	Do nothing	97.944	0
0	Do nothing	57.5171	0	1	Major Rehab	87.517061	300000	1	Preserve	92.5171	7500
0	Do nothing	72.5004	0	0	Do nothing	57.517061	0	1	Major Rehab	87.5171	13400
1	Preserve	100	4040	1	Preserve	100	4040	0	Do nothing	99.1825	0
1	Preserve	100	130	0	Do nothing	99.182549	0	0	Do nothing	97.944	0
0	Do nothing	72.5004	0	1	Minor Rehab	87.500353	20000	0	Do nothing	72.5004	0
0	Do nothing	0.41961	0	0	Do nothing	0	0	0	Do nothing	0	0
0	Do nothing	72.5004	0	1	Minor Rehab	87.500353	5000	1	Preserve	92.5004	500
1	Preserve	92.3895	8000	0	Do nothing	82.389479	0	0	Do nothing	72.5004	0
0	Do nothing	0.41961	0	0	Do nothing	0	0	0	Do nothing	0	0
1	Preserve	100	61000	0	Do nothing	99.182549	0	0	Do nothing	97.944	0

FIGURE 8.5
Excel set-up for infrastructure management.

AR	AS	AT	AU	AW	AX	AY	AZ	BB	BC	BD	BE
x	OWTreat 2018	PCI 2018	Cost 2018	x	OWTreat 2019	PCI 2019	Cost 2019	x	OWTreat 2020	PCI 2020	Cost 2020
0	Do nothing	97.944	0	1	Preserve	100	1000	1	Preserve	100	1000
1	Preserve	100	500	1	Preserve	100	500	1	Preserve	100	500
1	Preserve	100	51170	0	Do nothing	99.182549	0	0	Do nothing	97.944	0
0	Do nothing	99.1825	0	1	Preserve	100	500	1	Preserve	100	500
1	Preserve	93.9164	12500	1	Preserve	98.916403	12500	0	Do nothing	96.0675	0
1				0	Do nothing	99.182549	0	1	Preserve	100	13615
0	⇦ Decision Variables ⇨			0	Do nothing	97.944006	0	1	Preserve	100	24500
1				1	Preserve	100	38500	0	Do nothing	99.1825	0
1				1	Preserve	100	2500	1	Preserve	100	2500
1	Preserve	100	17585	0	Do nothing	99.182549	0	1	Preserve	100	17585
0	Do nothing	97.944	0	1	Preserve	100	2000	1	Preserve	100	2000
0	Do nothing	99.1825	0	1	Preserve	100	14500	1	Preserve	100	14500
1	Preserve	100	10000	1	Preserve	100	10000	1	Preserve	100	10000
0	Do nothing	97.944	0	1	Preserve	100	9000	1	Preserve	100	9000
1	Preserve	100	54000	1	Preserve	100	54000	0	Do nothing	99.1825	0
1	Preserve	100	500	0	Do nothing	99.182549	0	0	Do nothing	97.944	0
1	Preserve	100	3160	0	Do nothing	99.182549	0	0	Do nothing	97.944	0
0	Do nothing	99.1825	0	1	Preserve	100	10000	0	Do nothing	99.1825	0
	Objective ⇨	81.9068 Avg PCI	2960040 Total Cost	⇦ Constraint		82.955229 Avg PCI	2461230 Total Cost			83.7771 Avg PCI	2106175 Total Cost

FIGURE 8.6
Extended Excel set-up with objective and constraints.

for any number). Several elements from Excel are required for the solver: the decision variables (binary values of 1 or 0), the objective (say total network condition) and the constraints (say the total cost).

There are two ways one could approach the solution. If the solver is capable of handling several decision variables at a time, then one could run once for all decision variables at all periods. Otherwise, one would need to run year by year in chronological order. The previous example is part of the case study presented towards the end of this chapter. The commercial software 'Solver Risk Platform', an add-in to Excel, was used to solve this problem. This add-in to Excel is nothing but a more powerful solver add-in developed by the same creators of the commonly used solver in Excel. Its use requires the definition of the objective, the decision variables and their nature (binary) and the constraints. All this is achieved by simply selecting the corresponding cells in Excel that represent such elements.

8.6 Incorporation of Other Elements

This section is only intended to outline the way in which other elements could be incorporated into this problem. Elements such as environmental impact, safety or congestion could be added by expanding or adapting the framework presented here. Incorporating environmental impacts can be achieved by considering the amount of gas emissions from each type of MR treatment in a similar manner as that of cost. Each treatment will produce a given amount of unitary gas emissions per kilometre or square metre.

Incorporation of safety requires the development of a performance curve similar to that of condition but for safety. Instead of traffic load, we would use traffic volume and create a set of treatments for correcting safety deficiencies.

Adding congestion requires the inclusion of a performance model for the level of service, along with a set of treatments intended to improve traffic circulation. As seen in Figure 8.7, the improvements will reduce the travel time and hence increase the service area of proximal settlements and improve their access to hospitals. The dotted red line shows the initial coverage area, and the blue shaded area shows the coverage area after the improvements on the road network.

8.7 Extending the Decision over Several Criteria

This section is only intended to outline the way in which other criteria could be incorporated into this problem. So far, the problem was solved by looking into only one criterion. But in many occasions, we may need two or more elements. Think again of a pavement: an indicator for the condition of the surface would not reveal the decay of the structure. A similar case can be argued for a bridge where an indicator for the surface of the deck may be unrelated to the corrosion of the re-bars inside it. Hence, two indicators need to be used.

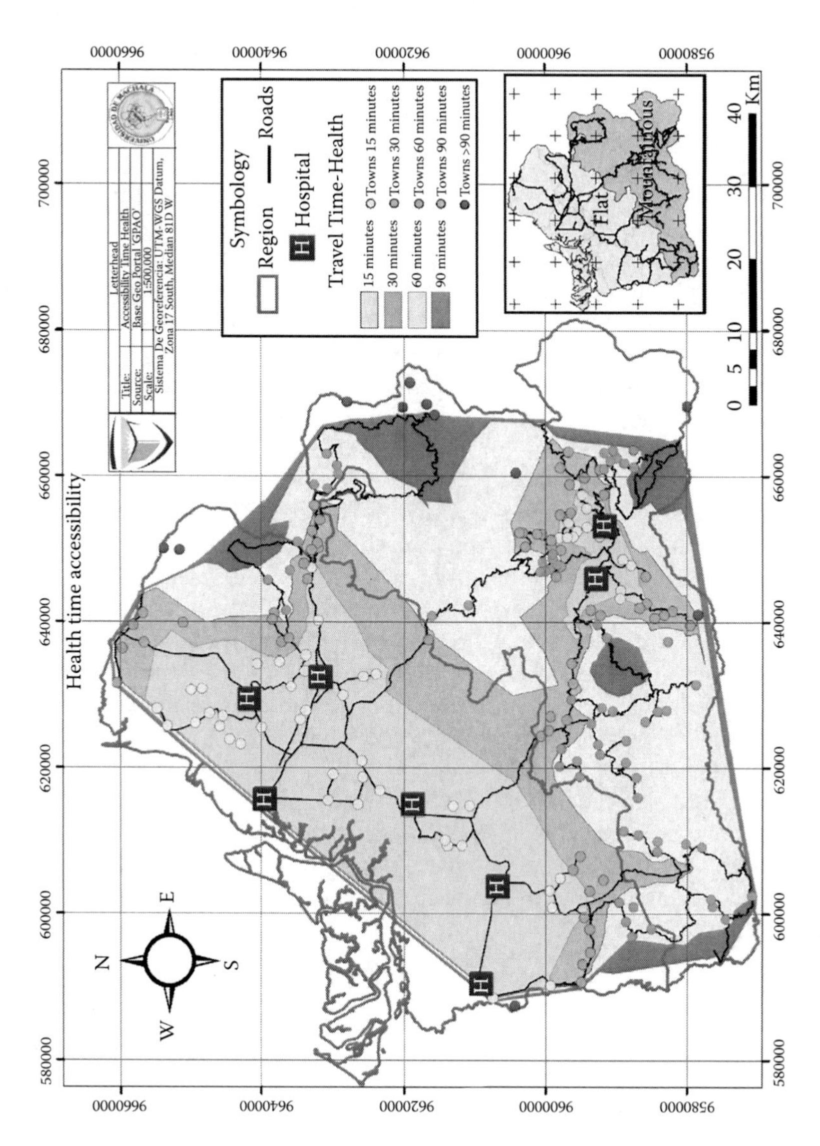

FIGURE 8.7
Travel time reduction strategy.

TABLE 8.6

Basic Intervention's Criteria

Intervention	Criteria	Structure/Surface
1. Crack sealing	$SAI > 80$ and $RUT < 0.5$	Good/good
2. MS	$SAI > 80$ and $RUT < 0.5$	Good/good
3. Overlay	$SAI > 50$ and $2 > RUT > 0.5$	Fair/fair
4. Reconstruction	$SAI < 50$ or $RUT > 2$	Poor/poor

When using two (or more) indicators, one needs to create independent performance curves for each indicator. The dynamic movement of the condition would need to be tracked on a per-indicator basis. For decisions, it is possible to separate per indicator, but in many instances we would want to combine them. For instance, you would seal cracks on a pavement if the level of rutting (RUT) is low and the pavement's structure adequacy index (SAI) is good. But if the structure is weak, there would be no point in sealing cracks if the entire structure is going to deteriorate rather fast. Table 8.6 shows sample ways to combine two criteria for pavement deterioration.

The previous structure had trouble in distinguishing between when to apply crack sealing and when to apply MS. This could be solved by extending to three criteria. Think now of the amount of linear cracking on the pavement. If crack sealing directly targets this issue, then perhaps it is worth having the decision criteria basis. An extended table could be built to include the percentage of the area being cracked. The overlay can be broken down into partial mill and overlay (3a) and full mill of the surface and overlay (3b). A more refined combination of criteria structure/surface can be seen in Table 8.7, where more appropriate combinations are matched to the corresponding criteria and intervention type.

It is interesting to see that the criteria for cheaper treatments involve more indicators to ensure good levels at both the structure and the surface. As we move down to more expensive treatments, the decision basis narrows and the structure component takes over a higher relevance, regardless of the value of surface indicators.

TABLE 8.7

Improved Intervention's Criteria

Intervention	Criteria	Structure/Surface
1	$SAI > 80, RUT < 0.5, CR < 20$	Good/good
2	$SAI > 80, RUT < 0.5, CR > 20$	Good/fair
3a	$SAI > 50, 2 > RUT > 0.5$	Fair/fair
3b	$SAI > 50, 3 > RUT > 2$	Fair/fair
4	$SAI < 50$	Poor/poor

8.8 Case Study

This example is an adaptation of a real project developed for a Provincial Government in Ecuador (El Oro) back in 2013–2014. The network comprised gravel (unpaved) and asphalt (paved) roads (Figure 8.8). Deterioration curves were developed and optimization scenarios were prepared.

The following paragraphs explain how these two elements were prepared. The notation for the average condition in the explanation that follows uses a three-letter code: the first letter denotes the average (u), the second letter indicates the year (4 = 2004, 6 = 2006) and the last letter indicates the condition (G = good, F = fair, P = poor, etc.).

For the deterioration curves, the procedure starts by assuming an apparent age (AGE1) of zero for the first IRI breakpoint (BP1) of 1.5 (m/km). This arbitrary assumption can be changed as needed. Second, the apparent age (AGE1) for the first pair of average good IRI points (u4G, u6G) was determined by finding the value of the age of the second break point (AGE2) that achieves the objective of separating the first pair of average IRI points (u4G, u6G) by a distance of 2 years (because of the time elapsed between

FIGURE 8.8
Gravel: left, coast; right, mountain.

condition surveys). The apparent age (AGE3) for the third breakpoint (BP3) uses the just established apparent age of the second breakpoint (AGE2) and finds the value of the corresponding age of the third break point (AGE3) that achieves a distance of 2 years between the second pair of average fair IRI points (u4F, u6F). This procedure continues in this fashion using the poor and very poor pairs of average condition until all apparent ages have been established. The following equation can be used for finding the apparent age of each break point, which is then modified to obtain the second equation that

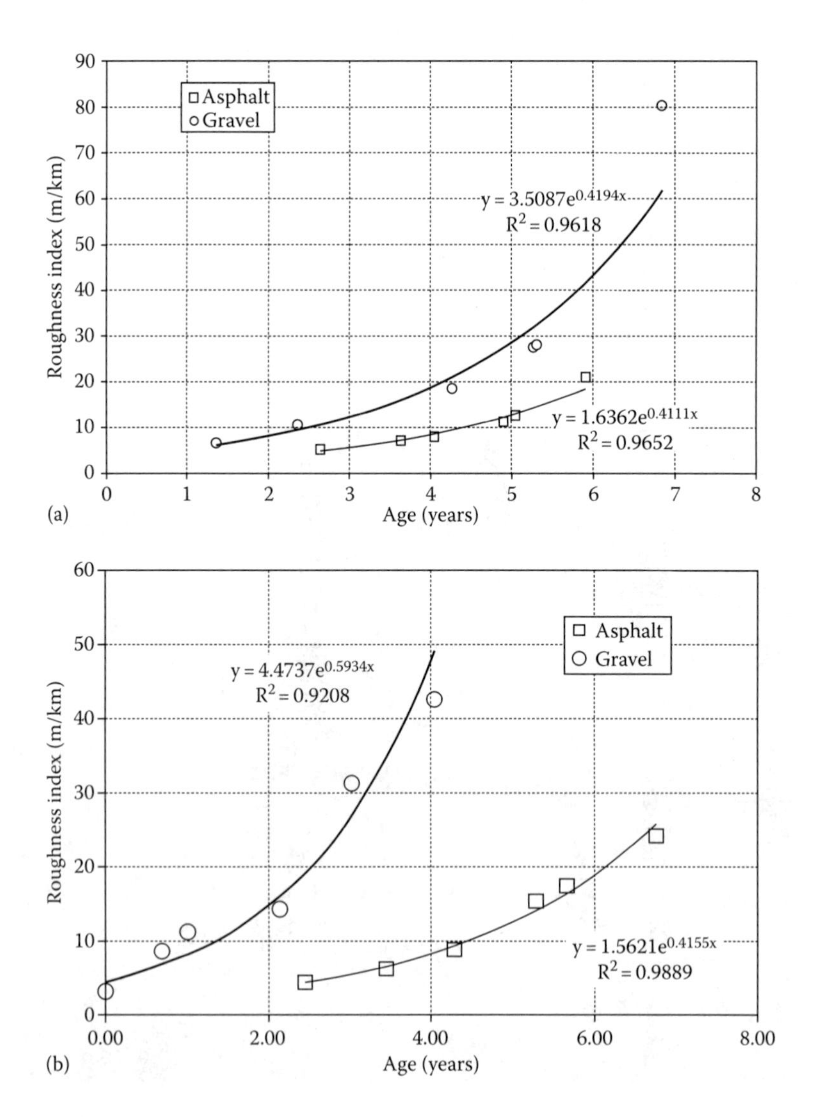

FIGURE 8.9
Deterioration: (a) mountain; (b) coast.

follows, which is an expression to obtain the individual values of age of each pair of average IRI points. Figure 8.9 shows the deterministic performance model for the mountain region and for the coast region for both asphalt and gravel.

$$\left(\frac{BP_n - u6i}{\frac{BP_n - BP_{n+1}}{AGE_{n+1} - AGE_n}}\right) - \left(\frac{BP_n - u4i}{\frac{BP_n - BP_{n+1}}{AGE_{n+1} - AGE_n}}\right) = 2$$

The apparent ages for the break points of the traffic intensity groups were used as a basis to assign the apparent ages for the entire database of observations. A direct formulation given by Equation 6 facilitates the direct computation of apparent ages for every observation. In the following equation, the variable x represents individual observations of IRI values at a certain point in time.

$$2\left(\frac{BP_n - BP_{n+1}}{x_4 - x_6}\right) + AGE_n = AGE_{n+1}$$

Apparent age intervals (operational windows) were matched to the IRI. Previous deterioration curves were used for this purpose. Only roads that improved in condition were used, that is, with values of IRI that were reduced between 2013 and 2014. The approach is very simple: first, one generates groups based on observed values of IRI during the year 2013, and then one counts how many segments moved across the intervals.

As seen in Table 8.8, for asphalt roads, those in good condition receiving an intervention extend their life span in 97% of chances by 2 years. Those 4 years old (apparent age) extend their life by 2 years in 7.41% and in 89.2% of the times in 4 years. Roads in poor condition (6 years old of apparent age) experience an extended life span of 2 years in 6.85% of times, 19.18% in 4 years and 69.86% in 6 or more years. Table 8.9 shows the intervention's effectiveness for gravel roads in the mountain regions.

Similarly, from Table 8.10, one observes for gravel roads in the coast region life span extensions of 2 years (93.87%) for those in good condition, of 2 years

TABLE 8.8

Intervention Effectiveness: Asphalt Roads

Age	0–2.46	2.47–5.61	5.62–12.77	12.78–29.08
0	100	0	0	0
2	97.05	2.95	0	0
4	89.2	7.41	3.4	0
6	69.86	19.18	6.85	4.11

TABLE 8.9

Intervention Effectiveness: Gravel at Mountain Regions

Age	0–8.11	8.12–18.78	18.79–43.51	43.52–100.51
0	100	0	0	0
2	25	75	0	0
4	9	36	55	0
6	2	4	42	52

TABLE 8.10

Intervention Effectiveness: Gravel in Coast Regions

Age	0–8.11	8.12–18.78	18.79–43.51	43.52–100.51
0	100	0	0	0
2	93.87	6.13	0	0
4	94.46	4.23	1.3	0
6	94.78	5.22	0	0

TABLE 8.11

Operational Windows and Cost

Condition	Gravel	Asphalt	Intervention (Asphalt/Gravel)	Cost ($/m^2)
Very good	NA	0–1.8	NA/crack sealing	5
Good	0–5	1.8–2.7	Partial/MS	30
Fair	5–10	2.7–5	Integral/mill overlay	50
Poor	15+	5–10	Preserve/reconstruct	300
Very poor	NA	10+	NA/reconstruction	300

(4.23%) and 4 years (94.46%) for those in fair condition and of 6 years (94.78%) for those in poor condition.

It is important to mention that the condition intervals and apparent ages for gravel and asphalt are measured in different scales and that gravel deteriorates much faster that paved roads.

Operational windows were established to replicate the local experience in terms of cost. Table 8.11 shows the operational windows developed for this case, which were used to restrict the options for the decision-making.

8.8.1 Scenario 1: Maximize Condition

The first scenario used as objective the maximization of asphalt road's surface condition with a maximum budget of US $3 million. Its results in terms of progression of condition across time and expenditure are shown in Figure 8.10 for asphalt roads and in Figure 8.11 for gravel roads.

As one can see, pavement fell from around 86 PCI to about 82 points (scale 100) and then recovered back after 6 years of full investment of US $3 million. Subsequently, after year 2021, the budget needed for MR of asphalt pavements fell around US $1.5 million due to the achievement of a network in very good condition levels of about 87 points. This means that about 50% of the budget (US $1.5 million) was released for other government activities.

Gravel roads with an average annual investment of US $8 million moved from 75 points of condition to about 93 point by the end of 20 years of investment. It is possible that with a smaller budget they could have achieved and remained at acceptable levels of condition. For this reason, a minimization scenario was run with the goal of achieving 75 points of condition for gravel roads and 80 points of condition for asphalt roads, as shown in Figure 8.12.

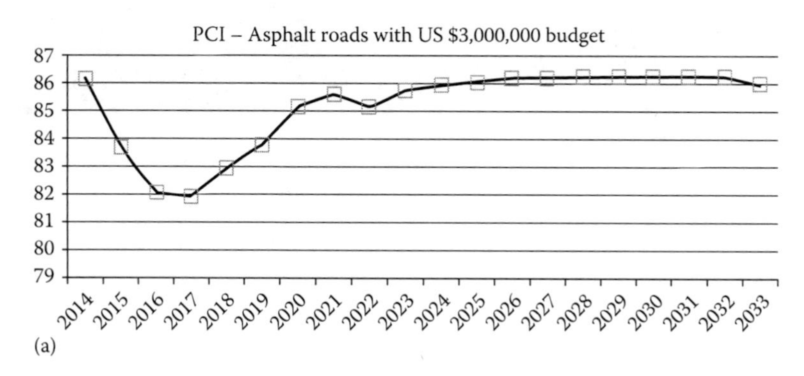

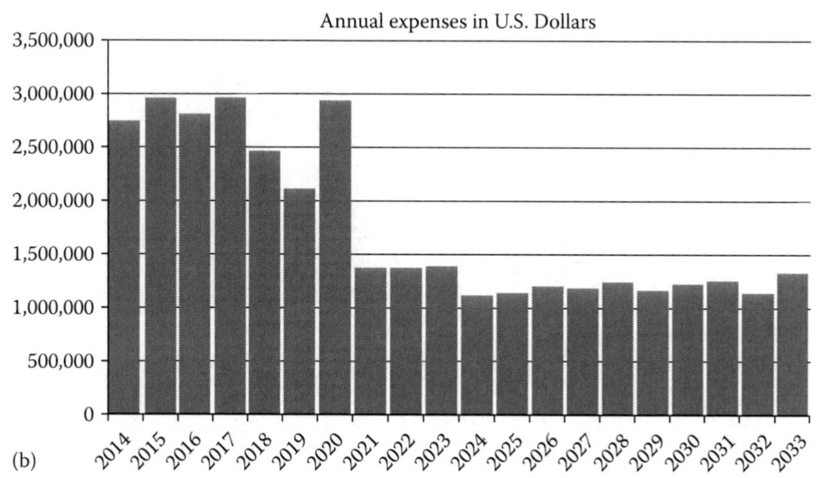

FIGURE 8.10
Scenario 1: (a) condition; (b) cost.

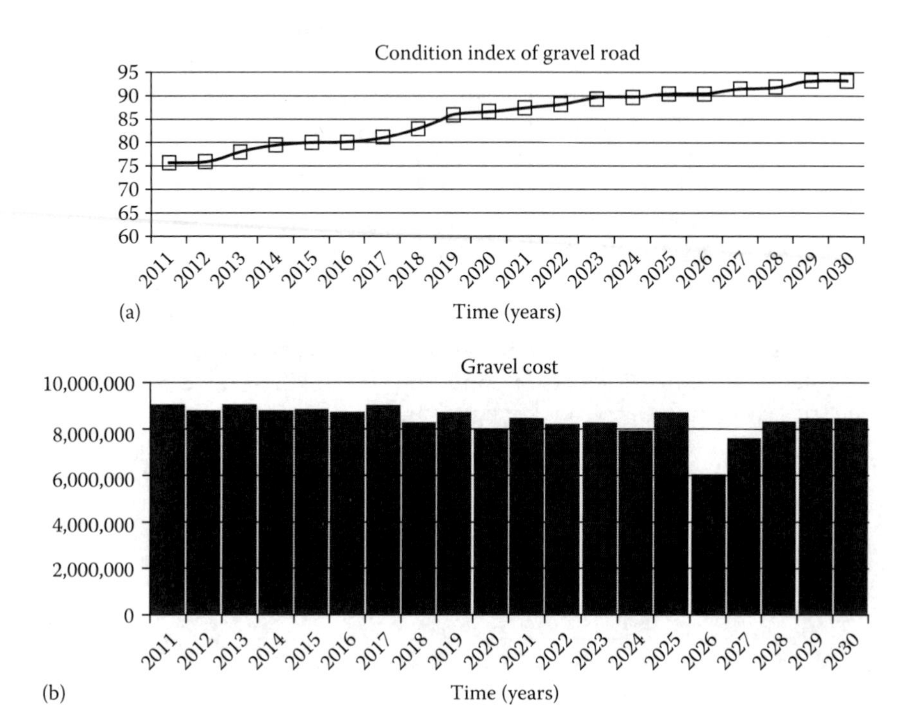

FIGURE 8.11
Scenario 1: (a) condition; (b) cost.

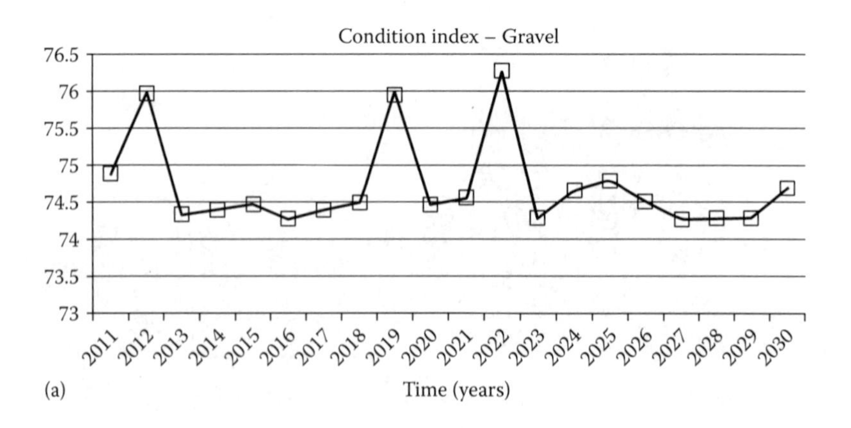

FIGURE 8.12
Gravel: (a) condition.

(Continued)

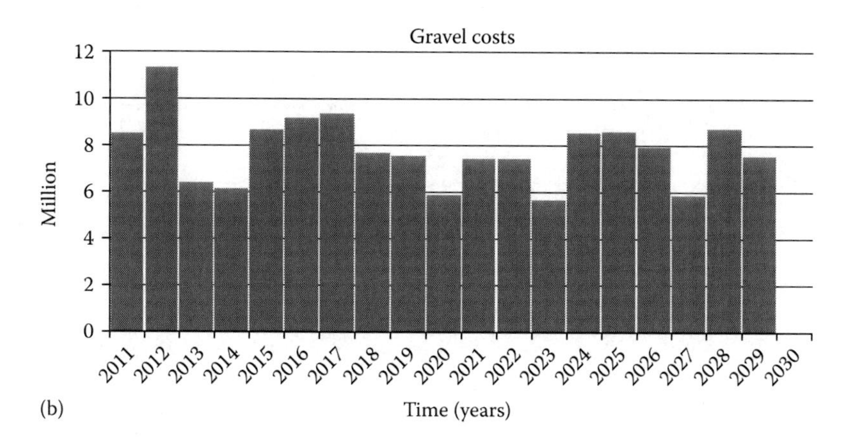

FIGURE 8.12 ***(Continued)***
Gravel: (b) cost.

8.8.2 Scenario 2: Minimize Costs and Reach Target Condition Level

This scenario fixed gravel road's condition to 75 points, and it was determined how much money was annually required to achieve it. It was found that US $7 million of annual budget was required to reach and remain at about 75 points of condition (Figure 8.11). For gravel, an average budget of US $9 million was required to hold levels of condition around 75 points (Figure 8.12).

In the case of asphalt, large variations of annual budget were observed (Figure 8.13), emulating and confirming the negative impacts of a poor first approach in which managers wait the fall of the network below 80 points. It is better to have constant levels of budget and maximize the condition (Figure 8.11).

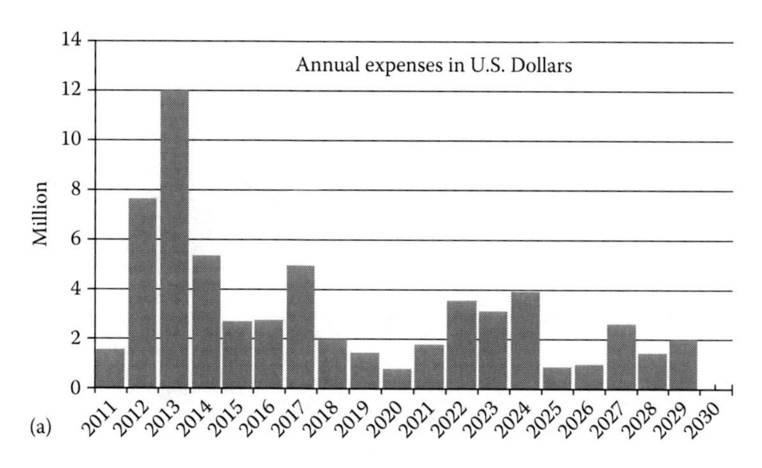

FIGURE 8.13
Asphalt: (a) condition.

(Continued)

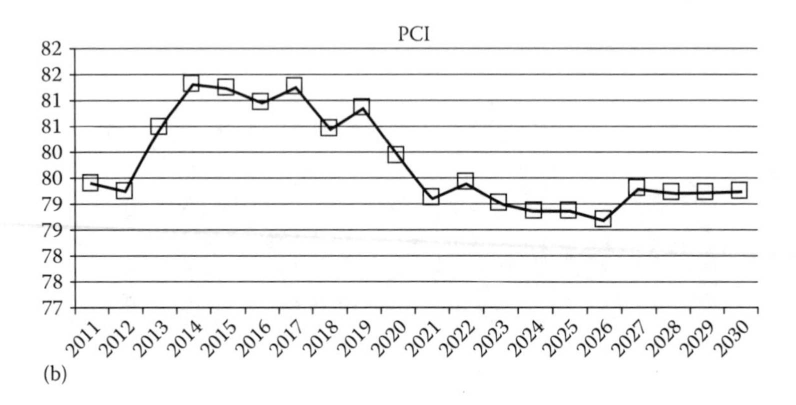

FIGURE 8.13 (*Continued*)
Asphalt: (b) cost.

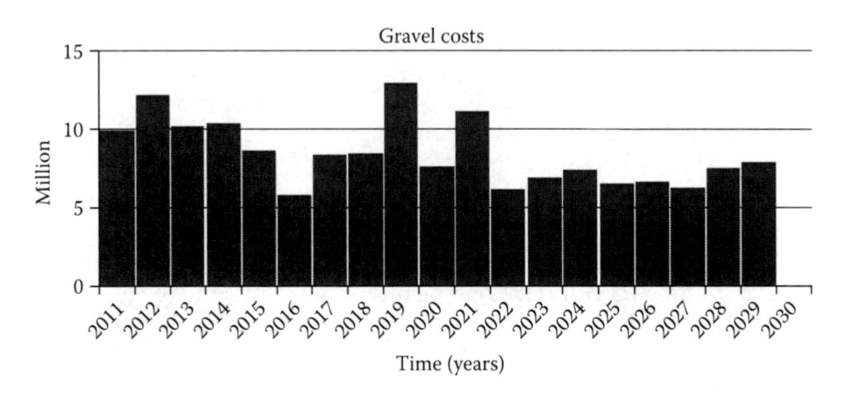

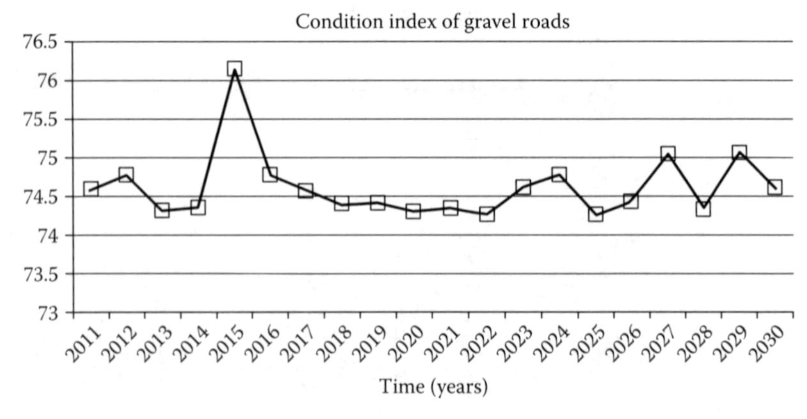

FIGURE 8.14
Gravel: Costs and condition.

8.8.3 Scenario 3: Minimize Environmental Impact

The third scenario tried to reach a minimum level of gas emissions (Figure 8.14) to reduce the effect of greenhouse gases (CO_2 equivalent). For this, the objective was the minimization of equivalent emissions.

For the case of pavement condition, its level remained between 79 and 80 points, while that of cost was very variable and environment impact decreased gradually in concordance with the wished goal (Figure 8.15).

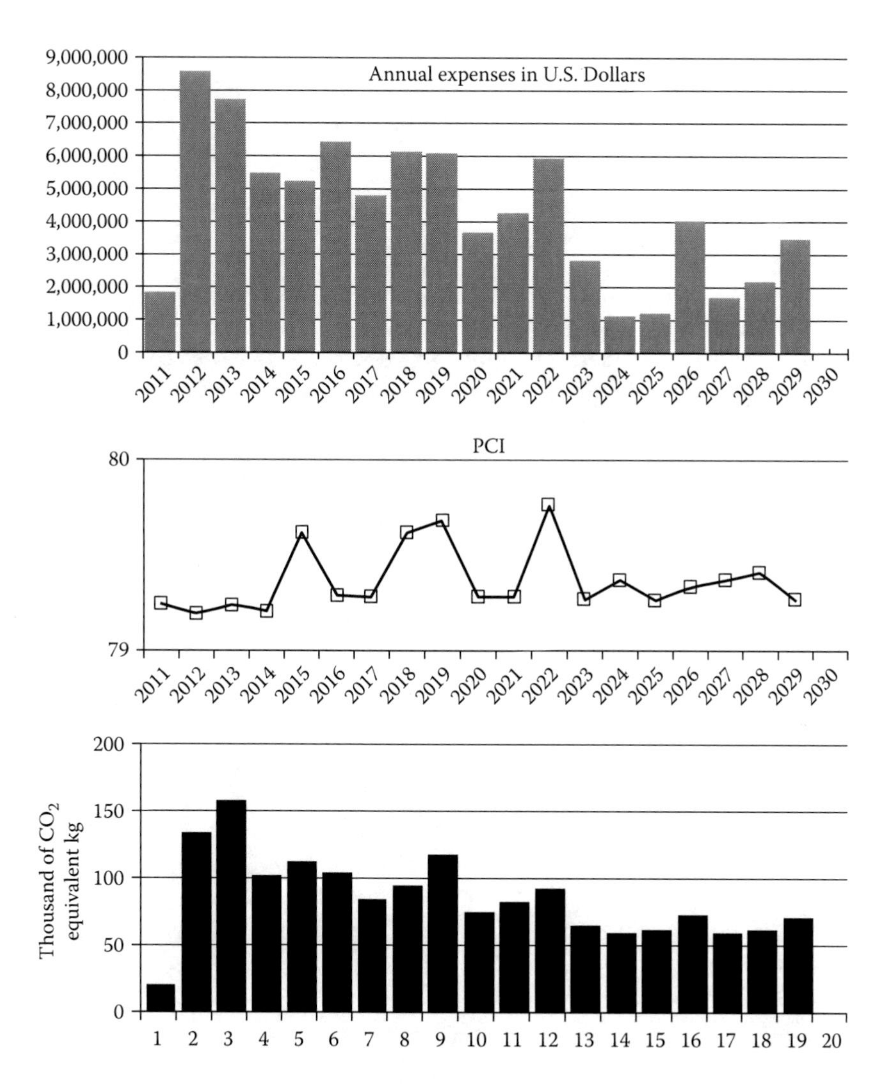

FIGURE 8.15
Pavement condition across time.

From this particular case study, one can conclude that treatment effectiveness for mountain regions showed very poor capacity of interventions to improve their condition. It is suspected that this is due to the use of local pits with large aggregates, which results in rough surfaces. On the contrary, the effectiveness at the coast showed very good values mostly due to the presence of more sand and better granularity of the materials.

Exercises

The exercises in this chapter are much longer and demanding than those in the previous chapters; they could take some days to be solved. This is due to the nature of the problem in hand. It is advisable to use the support of software (Excel, MATLAB) to solve them.

1. Terms of reference
   Objective: The new French Department of Transportation has asked you to analyse the assets in a small area of the province that will be put up for tender to be managed under a public–private partnership for the next 20 years. In order to evaluate the bids coming in from the various consortia, you will need to predict the rehabilitation needs of the road network in that area and project a minimum cost for maintaining the status quo condition, that is, do not let the average condition of the overall system decline. In order to ensure that the level of activity provides a sustainable road network, use a 20-year planning horizon.

Answer the following questions:

Is it reasonable to ask for non-declining condition given the current state of the highway network?

What would be a better management goal for the network?

What is the overall cost for the first 4 years for the non-declining scenario?

What is the overall cost for the first 4 years for chipseal in the non-declining scenario?

What is the best condition you can achieve with X million dollars (I suggest you use several values here; you can start with 5 and increase until you find a satisfactory solution)?

Inventory: The figures attached at the end of the enunciation show the inventory of segments.

Asset dynamics

Pavement treatment types

1. Minor rehab (mill and fill or thin overlay)
   a. Must have an SDI of > 75
   b. Must have an IRI of < 2.5
   c. Must have high strength
2. Major rehab (mill, base and fill or thick overlay)
   a. Must have an SDI of > 50
   b. Must have an IRI of < 3.5
3. Reconstruction (pulverize and rebuild)
   a. This can be done at any time regardless of the condition

Costs (note: Ignore inflation or discount rate.)

Minor rehab US $150,000/km
Major rehab US $300,000/km
Reconstruct US $350,000/km

Chipseal treatments

1. Level and reseal.
   a. Must have an SDI of > 50
2. Pulverize and seal.
   a. This can be done at any time regardless of condition but probably would not be considered for the first 10 years after the last age altering treatment

Costs

Level and reseal, US$ 30,000/km
Pulverize and seal, US$ 60,000/km

Deterioration: New asphalt regardless of treatment has a beginning IRI of 1.3 and degrades at 0.16 per year. New asphalt also starts at an SDI of 100 and declines by 3 per year (Figure 8.16).

New chipseal has a reset SDI value of 100 and declines by 4 per year.

The following set of figures show also the deterioration curves. Figure 8.17 illustrates the actual number behind the curves, while Figure 8.18 displays the actual curves.

2. The Ministry of Transportation wants to schedule interventions to improve pavement surface condition and road safety. Potential for improvement (PFI) indicates the difference between the observed and

Surface	Strength	Age	Length	Surface	Strength	Age	Length	Surface	Strength	Age	Length	Surface	Strength	Age	Length	Surface	Strength	Age	Length	Surface	Strength	Age	Length	Surface	Strength	Age	Length
AR	HS	1	7.006	AR	HS	3	125.454	AR	HS	5	24.164	AR	HS	7	7.135	AR	LS	8	1.102	CSR	HS	3	49.524	CSR	HS	3	55.555
AR	HS	2	17.199	AR	HS	4	72.18	AR	HS	6	57.746	AR	HS	1	22.21	AR	LS	9	2.755	CSR	HS	4	14.078	CSR	HS	4	0.96
AR	HS	3	16.855	AR	HS	5	90.928	AR	HS	7	26.119	AR	HS	2	104.947	AR	LS	8	10.812	CSR	HS	5	26.387	CSR	HS	5	37.772
AR	HS	4	5.302	AR	HS	6	30.203	AR	HS	1	18.564	AR	HS	3	298.115	AR	LS	8	5.633	CSR	HS	1	5.299	CSR	HS	7	0.46
AR	HS	5	5.248	AR	HS	7	19.334	AR	HS	2	46.378	AR	HS	4	117.221	AR	LS	9	2.415	CSR	HS	2	48.751	CSR	HS	1	23.947
AR	HS	6	37.972	AR	HS	1	32.718	AR	HS	3	92.346	AR	HS	5	82.875	AR	LS	8	1.384	CSR	HS	3	27.66	CSR	HS	2	67.229
AR	HS	7	8.746	AR	HS	2	137.136	AR	HS	4	56.551	AR	HS	6	30.052	AR	LS	9	4.45	CSR	HS	4	3.836	CSR	HS	3	69.129
AR	HS	1	45.256	AR	HS	3	180.319	AR	HS	5	53.184	AR	HS	7	40.905	AR	LS	8	19.218	CSR	HS	5	17.968	CSR	HS	4	15.656
AR	HS	2	39.83	AR	HS	4	15.567	AR	HS	6	2.385	AR	HS	1	49.979	AR	LS	8	33.121	CSR	HS	6	25.749	CSR	HS	5	9.655
AR	HS	3	60.297	AR	HS	5	31.529	AR	HS	7	3.152	AR	HS	2	165.783	AR	LS	9	13.526	CSR	HS	7	9.002	CSR	HS	1	7.1
AR	HS	4	52.298	AR	HS	6	9.865	AR	HS	1	5.123	AR	HS	3	161.436	AR	LS	8	19.521	CSR	HS	1	48.444	CSR	HS	2	3.813
AR	HS	5	4.062	AR	HS	7	14.212	AR	HS	2	17.125	AR	HS	4	49.745	AR	LS	9	6.103	CSR	HS	2	196.576	CSR	HS	3	6.125
AR	HS	6	20.777	AR	HS	1	58.73	AR	HS	3	87.89	AR	HS	5	39.884	CSR	HS	3	6.036	CSR	HS	3	91.644	CSR	HS	4	10.125
AR	HS	7	10.353	AR	HS	2	84.057	AR	HS	4	51.534	AR	HS	6	0.791	CSR	LS	8	0.49	CSR	HS	4	56.081	CSR	HS	5	9.971
AR	HS	1	9.422	AR	HS	3	112.237	AR	HS	5	8.423	AR	HS	7	49.207	CSR	LS	9	0.17	CSR	HS	5	7.022	CSR	HS	6	10.262
AR	HS	2	20.357	AR	HS	4	14.628	AR	HS	6	1.15	AR	LS	8	18.404	CSR	LS	9	0.429	CSR	HS	7	3.689	CSR	HS	7	2.37
AR	HS	3	114.278	AR	HS	5	13.617	AR	HS	7	2.688	AR	LS	9	2.745	CSR	HS	2	38.53	CSR	HS	1	10.286	CSR	HS	1	13.938
AR	HS	4	50.596	AR	HS	6	13.709	AR	HS	1	6.77	AR	LS	9	1.61	CSR	HS	3	53.01	CSR	HS	2	22.337	CSR	HS	2	54.608
AR	HS	5	38.547	AR	HS	7	9.513	AR	HS	2	97.957	AR	LS	10	1.21	CSR	HS	1	52.131	CSR	HS	3	20.919	CSR	HS	3	9.126
AR	HS	6	9.73	AR	HS	1	29.876	AR	HS	3	79.918	AR	LS	8	30.481	CSR	HS	2	91.428	CSR	HS	4	6.412	CSR	HS	4	0.135
AR	HS	7	3.214	AR	HS	2	49.669	AR	HS	4	18.922	AR	LS	9	3.507	CSR	HS	3	137.015	CSR	HS	5	1.031	CSR	HS	5	5.748
AR	HS	1	28.553	AR	HS	3	117.673	AR	HS	5	25.461	AR	LS	10	2.461	CSR	HS	4	33.44	CSR	HS	1	39.825	CSR	HS	2	42.167
AR	HS	2	70.815	AR	HS	4	61.106	AR	HS	6	30.667	AR	LS	8	18.398	CSR	HS	5	59.52	CSR	HS	2	82.297	CSR	HS	3	78.003
AR	HS	3	156.198	AR	HS	5	59.583	AR	HS	7	11.483	AR	LS	9	13.838	CSR	HS	1	22.1	CSR	HS	3	150.182	CSR	HS	4	44.596
AR	HS	4	74.129	AR	HS	6	59.502	AR	HS	1	15.382	AR	LS	8	15.336	CSR	HS	2	82.184	CSR	HS	4	31.246	CSR	HS	5	18.479
AR	HS	5	96.728	AR	HS	7	17.065	AR	HS	2	18.413	AR	LS	9	3.239	CSR	HS	3	117.743	CSR	HS	5	29.633	CSR	HS	6	7.047
AR	HS	6	32.996	AR	HS	1	1.467	AR	HS	3	75.535	AR	LS	10	3.865	CSR	HS	4	47.502	CSR	HS	6	4.297	CSR	HS	1	18.471
AR	HS	7	37.074	AR	HS	2	163.974	AR	HS	4	45.912	AR	LS	11	1.641	CSR	HS	5	7.744	CSR	HS	7	1.439	CSR	HS	2	37.131
AR	HS	1	17.553	AR	HS	3	90.335	AR	HS	5	22.93	AR	LS	8	14.007	CSR	HS	1	24.836	CSR	HS	1	24.437	CSR	HS	3	65.404
AR	HS	2	36.838	AR	HS	4	32.211	AR	HS	6	3.13	AR	LS	9	1.577	CSR	HS	2	63.647	CSR	HS	2	41.508	CSR	HS	5	2.664

FIGURE 8.16
Exercise 1.

Asphalt			Chipseal	
_age	YIRI	YSDI	_age	YSDI
1	1.3	100	1	100
2	1.46	97	2	96
3	1.62	94	3	92
4	1.78	91	4	88
5	1.94	88	5	84
6	2.1	85	6	80
7	2.26	82	7	76
8	2.42	79	8	72
9	2.58	76	9	68
10	2.74	73	10	64
11	2.9	70	11	60
12	3.06	67	12	56
13	3.22	64	13	52
14	3.38	61	14	48
15	3.54	58	15	44
16	3.7	55	16	40
17	3.86	52	17	36
18	4.02	49	18	32
19	4.18	46	19	28
20	4.34	43	20	24
21	4.5	40	21	20
22	4.66	37	22	16
23	4.82	34	23	12
24	4.98	31	24	8
25	5.14	28	25	4
26	5.3	25	26	0
27	5.46	22	27	0
28	5.62	19	28	0
29	5.78	16	29	0
30	5.94	13	30	0
31	6.1	10	31	0
32	6.26	7	32	0
33	6.42	4	33	0
34	6.58	1	34	0
35	6.74	0	35	0

FIGURE 8.17
Performance numbers.

predicted number of cars leaving the road (slippery circumstances). To avoid cars leaving the road and colliding with fixed objects, you can use a guardrail (a metallic beam) or could resurface (however, resurfacing only partially fixes the situation and hence it only gives its reduction on two points of PFI).

You want to minimize the PFI (in order to improve road safety) and to maximize the PCI. The drop in PCI happens only if you apply no treatment. If you apply an intervention (treatment), then PCI will increase by the amount given in Table 2 (effectiveness). All decisions are based on current values of PCI (today). Next year, the value is the result of today's plus the treatment effectiveness or the deterioration (decay) (Figure 8.19).

a. Prepare either one (unified) or two decision trees (or enumeration process) like the one solved in class; do it for 1 year only but for both condition and safety.

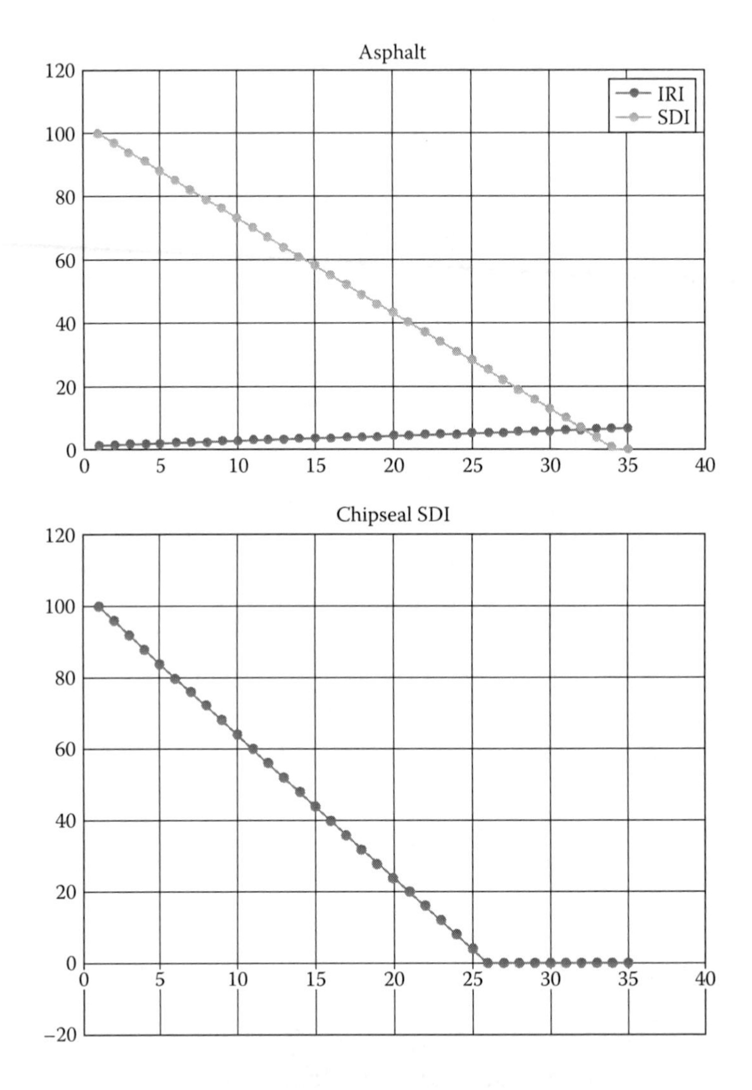

FIGURE 8.18
Performance curves.

b. Enumerate all possibilities (yes, all of them) and write down their cost, PFI and PCI values.

c. How much annual budget do you need for a 1-year analysis period in order to achieve at least average network PCI of 85 and PFI less than 2 points

d. What levels of PFI and PCI are the best you can achieve if the annual budget is US $500,000.

Table 1. Current Segments on the Network

Segment	Length (km)	Collisions/ km	SPF (Predicted)	PCI Deterioration	PCI Today	PFI Today	Safety Decay (Increase in PFI)
s1	3	10	8	10 points/year	70	2	1 point/year
s2	2	15	10	20 points/year	90	5	3 points/year
s3	1	20	7	10 points/year	50	13	1 point/year

Table 2. Treatment/Interventions Effectiveness

Treatments	Applicable If	Cost ($/km)	Emissions	Effectiveness	
				PCI	**PFI**
Resurface	80 ≥ PCI ≥ 60 or PFI < 8	200,000	40	+20 points	−2 points
Reconstruction	PCI ≤ 59	500,000	100	PCI = 100	N/A
Guardrail	PFI >8	40,000	10	N/A	PFI = 0

FIGURE 8.19
Exercise 2.

9

Uncertainty

9.1 Introduction

This chapter discusses issues related to uncertainty and the use of probability theory to incorporate uncertainty into decision making. We will use often the concept of value. Value is a monetary expression that captures the desirability associated with a given event or strategy. It is not the actual amount of money. So the value associated with a strategy is given by another weighted average that sums up positive and negative outcomes.

The simplest case in which you have probably faced uncertainty is that of a lottery. Suppose the reward is \$1 million and the ticket price is \$100; however, there are other 100,000 tickets, so is there any way to determine whether this is a good deal (formally called arbitrage)? Simply multiply the odds of winning by the reward amount and subtract the price of the ticket. In this case, the odds of winning and the value of the lottery are as follows:

$$P(win) = \frac{1}{100{,}000} = 0.00001$$

$$0.00001 * 1{,}000{,}000 - 100 = -90$$

As you can see, there is a negative value associated with playing this lottery. What if the price drops to \$1? Then you should definitely buy one ticket. Actually, any price strictly below \$10 would give us a net positive value associated with this lottery. Engineers face similar situations in their professional practice, and it is useful to know how to handle them in order to still make decisions.

One of the most common situations where the decision maker faces some degree of uncertainty is when he or she needs to submit a bid, because the price of the bid plays a very significant role. However, a company can bid a high price when the nature of the works is very specialized and there is no competition.

Often, other elements are considered, such as the contractors' record of completed similar works, their construction capacity (measured in machinery, personnel, etc.) or their experience. Winning a bid comes down to a final comparison of the offers made by the bidders.

A company may wonder whether for a given job they could take some risk and submit a high price, and whether for another one they should be more conservative and bid a low price, just to keep the employees busy and the business rolling but perhaps not making much profit. Let's look at a common strategy to measure the value associated with events.

9.2 Measuring Desirability: Value Approach

Consider there is a tender to compete for the repairs that a road needs. Assume that the degree of damage can be estimated by an inspection allowed during the preparation of a bid. The inspector identified the odds of having a road in very poor (48%), poor (32%) or fair (20%) conditions.

This is based on inspector's expert criteria (perhaps with the support of some machine), and it signifies that if the company bids the works for 1 million dollars, the repairs associated with such bid if the road was really fair (300,000) will give them a profit of 700,000, but if the road was poor, the repairs will cost more (600,000) and the profit will shrink to 400,000, and if the road was very poor, then the works will cost a lot (1,300,000) and the company may potentially lose 300,000.

This is shown in Table 9.1. If the cost of the preparing the bid is US $125,000, should the company apply for it? As you can see, you should not participate in this tender under this circumstance. You have a negative expected value of 124,000 − 125,000 = −1,000.

Now perhaps the manager could figure out that they must charge a higher price for their works. I will leave this event as Exercise 9.1 at the end of this chapter.

A historical record of previous contracts with the government may be available to a company. Such record will show the company, out of those

TABLE 9.1

Historical Profit and Its Probability

Bid Price	Repairs	Profit	*P*(*Profit*) or *P*(*Condition*)	Value
1,000,000	300,000	700,000	20% Fair	140,000
1,000,000	600,000	400,000	32% Poor	128,000
1,000,000	−1,300,000	−300,000	48% Very poor	−144,000
Cost =	125,000		Expected profit =	124,000

TABLE 9.2

Road Repair

Bid Price	Profit	*P*(*Profit*)%	*E*(*Profit*)
High	800,000	20	160,000
High	500,000	32	160,000
High	−100,000	48	−48,000
		Expected profit =	272,000
Low	400,000	60	240,000
Low	200,000	30	60,000
Low	−500,000	10	−50,000
		Expected profit =	250,000

projects awarded to them, what profit they made. For instance, a company may observe the probabilities associated with different levels of profit depending on whether the bid price is low or high.

This makes sense since the company faces the uncertainty of many kinds, such as soil-bearing capacity, the weather during the time of construction, the cost of mitigation strategies during the construction of the project and so on. As you can imagine, the odds of being awarded the project are higher when the price is low (80%) and smaller when the price is high (20%). The preparation of the bid will also incur a cost (40,000).

The question is: Should the company submit a high bid price or a low bid price? We proceed to obtain the expected profit for a high price and for a low price, as illustrated in Table 9.2.

When we look at the expected profit, it is clear that a high bid has a higher value and should be preferable. The problem is that the likelihood of getting that contract is only 20%, and yet you have to spend $40,000 for its preparation. So when you multiply the expected profit by the likelihood you get and adjusted value that reduces to $272{,}000 * 0.20 = 54{,}400$ to this, you need to reduce the cost of the bid: $54{,}400 - 40{,}000 = 14{,}400$.

Again, you must bear in mind that this is a monetary measure of desirability and does not represent profit any more. For the case of the low bid, you have $250{,}000 * 0.8 = 200{,}000$ minus the cost of preparing the bid (40,000), which leads us to a total value of 160,000.

This immediately tells you that the desirability of going for a low bid is higher than a high bid. The calculations are shown in Table 9.3.

Now, assume you have the chance of bidding for another project in addition to the previous one, say the construction of a rail line. The problem is your resources only allow you to bid for either the road repair or the rail line. Let's assume the values for high and low bid for the rail line are as given in Table 9.4. Also, let's suppose you have better odds of getting a high bid, actually 40% odds, as there is scarcity of rail line constructors; the odds of getting

TABLE 9.3

Road Repair: Expected Profit and Value

Bid Price	Profit	*P(Profit)%*	*E(Profit)*	*P(Win)*	Value	Final
High	800,000	20	160,000			
High	500,000	32	160,000			
High	−100,000	48	−48,000			
			272,000	0.2	54,400	14,400
Low	400,000	60	240,000			
Low	200,000	30	60,000			
Low	−500,000	10	−50,000			
			250,000	0.8	200,000	160,000

TABLE 9.4

Rail Line: Expected Profit and Value

Bid Price	Profit	*P(Profit)(%)*	*E(Profit)*	*P(Win)*	Value	Final
High	700,000	40	280,000			
High	400,000	40	160,000			
High	−100,000	20	−20,000			
			320,000	0.4	128,000	88,000
Low	500,000	30	150,000			
Low	300,000	30	90,000			
Low	−200,000	40	−80,000			
			160,000	0.6	96,000	56,000

a low bid are 60%. Let's suppose the cost of preparing the bid for the rail is the same (US $40,000).

As you can see if the decision is made to go for the rail line, then you should shot it with a high price, because this results in a much higher value, that reflect the fact that you have good odds (40%) of getting positive profit (80%); meanwhile, the odds of getting a negative profit on the low bid are quite significant (40%).

Now, let's remember we have two options actually not only one, so we need to consider them both together: road repair or rail line, either high or low. Let's put them together in Table 9.5. It is clear that the more desirable choice turns out to be bidding for a road-repair and submit a low price. Note the low price has an 80% chances of being awarded which results in a high value after subtracting the cost of preparing the bid (40,000).

At this point, I should mention that some authors have applied to the cost the likelihood of not winning; however, my opinion is that such cost

TABLE 9.5
Road and Rail: Combined Values

Bid Price	*E*(*Profit*)	*P*(*Win*)	Value	Final
Road–high	272,000	0.2	54,400	14,400
Road–low	250,000	0.8	200,000	160,000
Rail–high	320,000	0.4	128,000	88,000
Rail–low	160,000	0.6	96,000	56,000

must always be subtracted because you do incur it. The information can be combined to produce a graphical representation of the problem.

The idea is to start the graph from the right and move backwards to the left. As you move, the tree branch size narrows down as you use probabilities to combine choices into fewer elements. We start with the expected profits for the high bid for a road (800,000, 500,000, −100,000) and combine them using the probabilities (20%, 32% and 48%).

Continue in a similar manner with the expected profit values for low bid for road (400,000, 200,000, −500,000) and place them vertically below the previous ones. Bring also the probabilities (60%, 30% and 10%) Repeat in the same way for rail for both high and low. Figure 9.1 shows the final decision tree highlighting in amber the selected path.

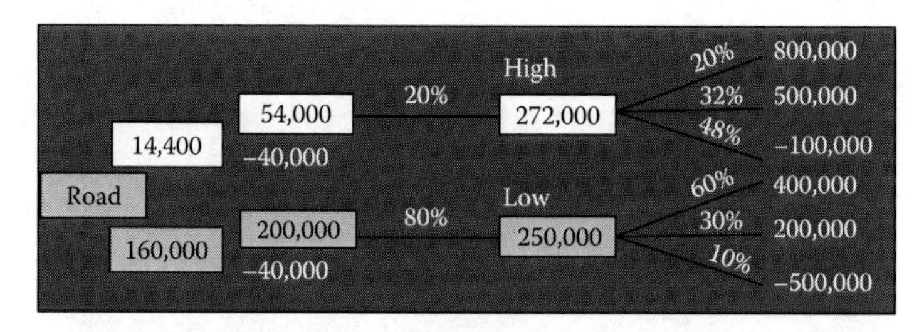

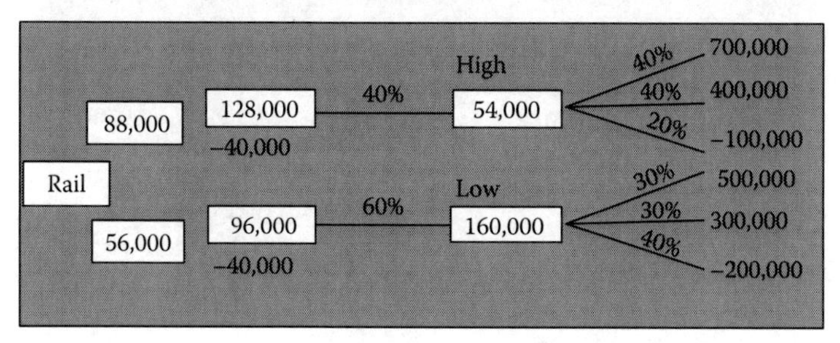

FIGURE 9.1
Decision tree.

9.3 Incorporating Experiments

Suppose now that before making the decision you have the choice of conducting an experiment which will give you some indication of the actual reality, that is, whether the soil capacity is high, the pavement structure is strong, the extent of repairs is minimum, the corrosion of the re-bars is negligible or not, and so on. This section explains the extension of decision trees that incorporate an extra branch, accounting for the possibility of conducting an experiment to learn the true state of our system. In Chapter 4, we learned to estimate marginal probabilities and use them to estimate the likelihood of an experiment of being correct under different states of our system. Let's reproduce Figure 4.9 here (we will call it, however, Figure 9.2).

Recall that we used a device called 'deflection meter' to estimate how strong a pavement structure is. The problem is the machine does not provide you with infallible results. Sixty-seven percent of the time, it detects a weak structure as such, 60% as fair and 69% as strong. How do we incorporate this into our decision making. Suppose you are submitting a bid for repairing a highway. You could simply choose not to conduct any experiment and simply select a full reconstruction, which will cost you $700,000 per lane-kilometre, or you can take the risk of simply overlaying the highway, which will cost you $200,000 per lane-kilometre. The problem is if you do the overlay, the structure may be weak and the highway may deteriorate very fast and your company will become liable for such issue. This could potentially signify that you will have to remove the overlay (cost of 100,000) and reconstruct the road (cost of 1,000,000). If the structure capacity is medium, your overlay may require some repairs (cost of 100,000); if the structure is strong,

		Experiment result: deflectometer result		
		Weak	Fair	Strong
Real state of nature: pavement strength	Low	**0.26/0.39 = 0.67**	**0.05/0.35 = 0.14**	**0.01/0.26 = 0.04**
	Medium	**0.08/0.39 = 0.20**	**0.21/0.35 = 0.60**	**0.07/0.26 = 0.27**
	High	**0.05/0.39 = 0.13**	**0.09/0.35 = 0.26**	**0.18/0.26 = 0.69**
Subtotals		**1**	**1**	**1**
Marginal probability of the experimental result		0.39	0.35	0.26

FIGURE 9.2
Conditional probabilities.

TABLE 9.6
Cost Breakdown

Real State	Overlay Cost	Repair Cost	Total Cost
Low	200,000	1,100,000	1,300,000
Medium	200,000	100,000	300,000
High	200,000	0	200,000

your overlay will be fine and no further repairs will be required. All these options are encapsulated in Table 9.6. Remember, you have to add to this the fact that you already spent $200,000 in the initial overlay.

Let's now bring in the probabilities. The way to combine all this information is by building a decision tree that first estimates the expected cost that we will incur if we decide to go for an overlay. We depart from the assumption that the pavement strength is weak, so we use the 13%, 20% and 67% as the associated likelihood of the machine identifying a high, medium or low pavement strength. We repeat for a fair strength and place the three-item branch underneath the weak one; we repeat again for the strong and place it vertically under the two before. Finally, we use the marginal probability that represents the odds of the instrument identifying a weak, fair or strong pavement. The value or desirability of conducting the experiment sums up to a cost of $582,650. We now compare this to the choice of not conducting any experiment and simply doing an overlay or a full reconstruction (Figure 9.3).

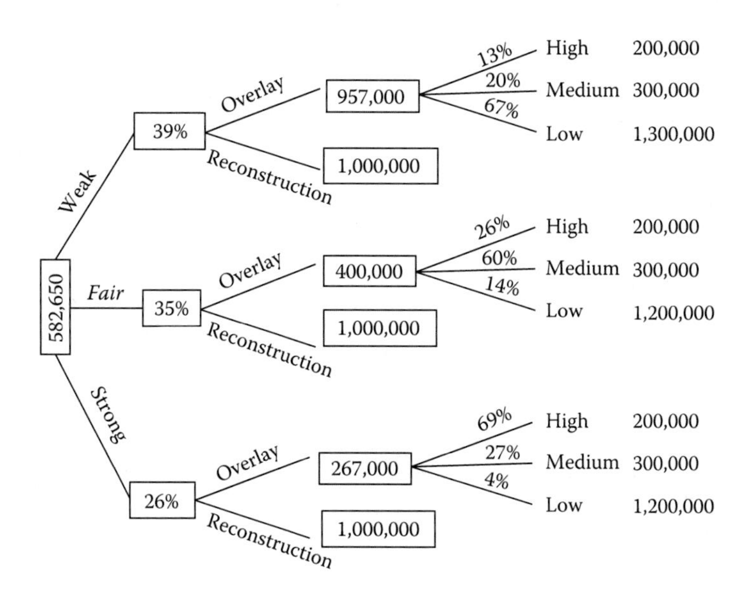

FIGURE 9.3
Experiment decision tree.

If we decide to a full reconstruction, it will cost us US$ 1 million, which is higher than the value or desirability of conducting an experiment, which is of 582,650, so it is cheaper to go for the experiment. You may wonder what if we go for the overlay. In this case, we need to create a decision tree for the overlay itself. The marginal probabilities for the real state of nature will be used in this event (Figure 9.4). So we use 32%, 36% and 32% for the high, medium and low; we repeat here Figure 9.5.

After incorporating these probabilities, the decision tree increases in size. We add new branches for the overlay and the reconstruction. In this case, we are looking for the cheapest value because what we have is a reflection of the cost. The cheapest way to go about this is by selecting an experiment; just

		Experiment result: deflectometer result				Marginal probability of the state of nature
		Weak	Fair	Strong	**Subtotals**	
Real state of nature: pavement strength	Low	**26**	**5**	**1**	32	0.32
	Medium	**8**	**21**	**7**	36	0.36
	High	**5**	**9**	**18**	32	0.32
Subtotals		39	35	26		
Marginal probability of the experimental result		0.39	0.35	0.26		

FIGURE 9.4
Marginal probabilities for the state of nature.

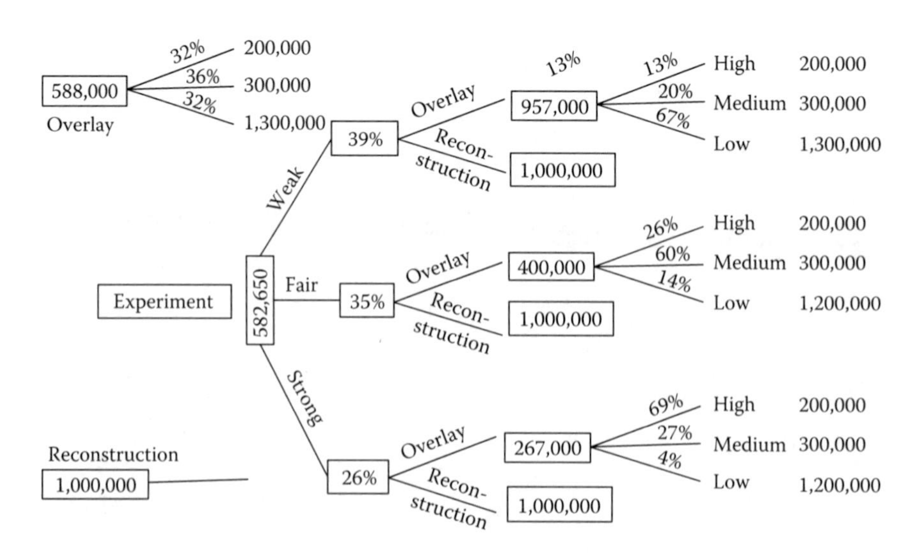

FIGURE 9.5
Final decision tree.

with a few dollars, the experiment is cheaper than going directly for the overlay. Notice that we have not assigned a monetary value to the experiment. This cost could increase beyond the cost value of going for an overlay, and hence there is a possibility that the experiment could be rejected.

9.4 Uncertainty in Coefficients

This section illustrates the process of estimating the probability distribution of the coefficients of any equation. Such estimation is always guided by a set of observations, typically called the data. This builds on the concepts introduced in Chapter 5. However, the reader must notice that in this section we will be able to get a clear picture of the uncertainty associated with each coefficient.

In Chapter 5, we were able to obtain only one value per coefficient, and you notice we did not worry about how much variation such coefficient could have. The variability associated with the response is covered in Section 9.5. In this section, we concentrate on the variability of the coefficients that accompany each explanatory variable.

In Chapter 5, we learnt to estimate the explanatory power of a group of coefficients in order to measure their ability to explain a given response. We also learned that the p-value associated with a value must be smaller than a given value (0.05 for 95%, 0.2 for 80%).

So, for instance, in Table 5.2 we had a given data set, and using a given functional form we were able to estimate the values of $\alpha = 0.609$, $\beta = 0.781$ and $\delta = 6.06$. However, you may wonder if the actual value of each coefficient exhibits a lot of variability.

We already learnt a way to see how sensitive the response is to each coefficient, but what if the value of one coefficient varies a lot and the value of another one does not? Would it be easier to simply assume the coefficient as fixed and concentrate our attention to characterize the reasons behind the large variability of the first one?

Besides, this will provide us with a very good indication of uncertainty, because it will tell us which element brings more variation to our analysis. Remember, in many occasions, that this is part of the actual process you will develop when trying to understand how a system works. The very same equations that you follow to estimate the amount of re-bars on a beam or a column are expressions that were born in this fashion.

Let's reproduce here Table 5.2 and call it Table 9.7, but this time we will use the open-source free software *OPENBUGS*, which can be freely downloaded from the Internet and installed in your computer. I won't show how to do it as it is very simple. *OpenBUGS* is the successor of *WinBUGS*, and you can alternatively install or use this one if you wish.

TABLE 9.7

Sample Data Deterioration

IRI	*ESALs*	*SNC*	*m*
0.18	60,000	3.96	0.04
0.45	122,000	3.96	0.06
0.67	186,000	3.96	0.08
0.54	251,000	3.96	0.04
0.95	317,000	3.96	0.06
1.08	384,000	3.96	0.08
0.83	454,000	3.96	0.04
1.46	524,000	3.96	0.06
1.48	596,000	3.96	0.08
1.17	669,000	3.96	0.04
1.89	743,000	3.96	0.06
1.96	819,000	3.96	0.08
0.88	897,000	3.96	0.02
1.82	976,000	3.96	0.04
1.67	1,057,000	3.96	0.05
2.66	1,139,000	3.96	0.07
1.30	1,223,000	3.96	0.02
1.43	1,309,000	3.96	0.03
2.40	1,396,000	3.96	0.05
2.39	1,486,000	3.96	0.04

We use either of these software packages to estimate the coefficients, with the assumption that they will be able to provide us with an estimation of uncertainty for the coefficients and for the response. They will actually draw the density function associated with each coefficient and also with the response. First, we concentrate our attention on the coefficients.

We need an equation that relates the response (*IRI*) with the causal factors ($x_1 = ESALs$, $x_2 = SNC$ and $x_3 = m$). As you will recall, the functional form we used was given by the following expression:

$$IRI = \frac{(m - index)^{\alpha}(ESALs)^{\beta}}{SN^{\delta}}$$

We will use the previous data set to explain how to set up *OPENBUGS*. You need to create two elements: (1) the data set and (2) the functional form. Creation of the data set is very straightforward. There are two possibilities, column format and row format. I will only explain the column format as it is the most common and natural one for an engineer. First, open your data set in Excel or any other similar software. Second, copy and paste into *Notepad* or any other *.txt* plain text editor. The first row corresponds to your labels and they will be extensively used in the creation of the functional form. I suggest you keep the names to no more than three letters where possible. Add [] at the end of each name, so, for instance, in our example we have *IRI*[] *ESAL*[] *SNC*[] *m*[]. One space of separation is enough, but you can have more; the software will read the separation as an indication of a new variable. At the very bottom below the last row, add a new row and write in all capital letters *END*, and then save the file as *.txt*. This is all you need to prepare your data.

For the preparation of the model, you need to identify a functional form that contains the variables and the response. We have done that already. Now, you need to define the mean expectation instead of the response. For this, we commonly use the notation *mu*, but you could use any word you wish really. We define the mean value of the response to be given by the functional form (equation) obtained before. The symbol for equal to is $< -$. When you want to express a power function, you use $pow(base, power)$, so, for instance, in our equation we have $IRI = \frac{(m-index)^{\alpha}(ESALs)^{\beta}}{SN^{\delta}}$, which will be expressed as $mu[i] = pow(m[i], alpha) * pow(ESALs[i], beta)/pow(SN[i], delta)$.

You probably notice that we have added to each variable the ending [*i*], whereas for our parameters or coefficients we did not. We also added it to the response. So we are missing an equation to state that the new variable *mu*[*i*] is equal to our old variable *IRI*. We do this by defining IRI as a random variable that follows a probabilistic distribution, for which we define the mean (*mu*[*i*]) and precision (*tau*). In our case, this takes the form of a normal distribution, and its notation is $IRI[i]\, follows\ dnorm(mu[i], tau)$. It is time now to define the precision (*tau*). It is nothing but 1 divided by the standard deviation squared.

Not only our response follows a probabilistic distribution, but our coefficients also follow the same. So we define α, β and δ similarly as before. Figure 9.6 presents the *OPENBUGS* model.

Note we have defined a loop for 20 observations indexed by *i*; then we have defined our response to follow a normal distribution with mean *mu*[*i*] and precision *tau*. The next line defines the functional form. A standard residual and its probability follow and then a new variable called predicted. These terms will not be used in this example. Most importantly, we find the definition of α, β and δ, all as normally distributed. Notice we are using as mean value the values previously found. A precision of 16 means a standard deviation of 0.25 or two standard deviations of 0.5 for the 95% confidence interval. This specification is called 'prior' and could be corrected if the data strongly support its bias.

In the next step, we open the database on *OPENBUGS* (do not close the model; you will need it very soon). We will then open several windows within the interface. On the model tab, open the *model specification* window. Highlight the word *model* on the *model window* and hit the *check model* at the bottom of the *specification tool*. You will see a message at the bottom of the window, which hopefully will read 'model is syntactically correct'. Highlight the label

```
ch9_1
# BOOK CESA, chapter 9.
model {
  for (i in 1:20) {
   IRI[i] ~ dnorm(mu[i],tau)  # adapted from Patterson and Attoh-Okine
  mu[i] <- pow(m[i],alpha)*pow(ESALs[i],beta)/pow(SN[i],delta)
   res[i] <- (IRI[i]-mu[i]/sigma)                    #standard residual
   p.res[i] <- phi(res[i])                          # accumulated std residual
  Y.pred[i] ~ dnorm(mu[i],tau)
#   P.pred[i]<-step(IRI[i]-Y.pred[i])   # indicator of whether observed > predicted
  }
alpha ~ dnorm(0.609,16)                          # Prior expectations informative
beta ~ dnorm(0.781,16)                           # 2sd = 0.5, i.e. intercept 0.281 to 1.281
delta ~ dnorm(6.06,16)
tau ~ dgamma(0.0001,0.0001)
sigma <- 1/sqrt(tau)
  }
list(alpha=0.7, beta=.9, delta=0.4, tau=0.0001)
list(alpha=0.5, beta=0.6, delta=0.8, tau=0.0001)
```

FIGURE 9.6
Final decision tree.

row from the data tab and hit on *load data* at the *model specification* window. At the bottom, you will see a message saying *data loaded*. The set-up is shown in Figure 9.7.

It is time to tell the software about the two departure points called chains. These are values of the coefficients we specify as initial points; first, change the number to match that of the chains you want to specify and then

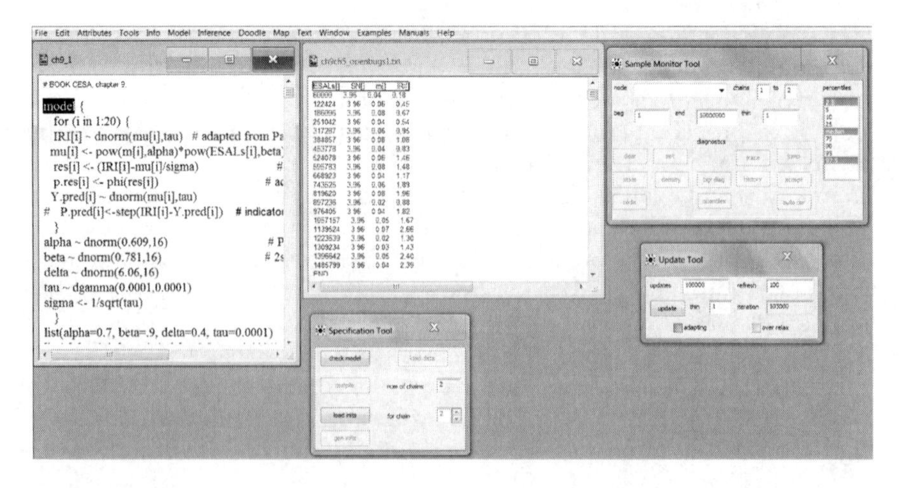

FIGURE 9.7
OPENBUGS interface.

highlight the word *list* from the first chain and then *load inits* from the model specification window. Repeat this for the second chain. Sometimes, some parameters need to be initialized, so hit the *gen inits* to complete the specification. Select *Samples* from the *Inference* tab, write on the *node* box the coefficients to be estimated one by one and hit *set* after each name has been input on the white box. You can also keep track of the mean if you need to. Next, select *update* from the *Model* tab, and specify the number of iterations to be run. Figure 9.7 illustrates all the windows used to specify the model.

The convergence of the system is important. One way to study this is by looking at the chains and observing how they progress from separate initial values to a common tight range. On the *sample monitor tool*, specify the beginning (*beg*) and ending (*end*) iterations to visualize chains within such range. Specify the coefficient you want to visualize on the *node* box. Hit on the bottom named *History*; this will plot the chains. You may need to change the beginning (*beg*) and ending (*end*) values (Figure 9.8).

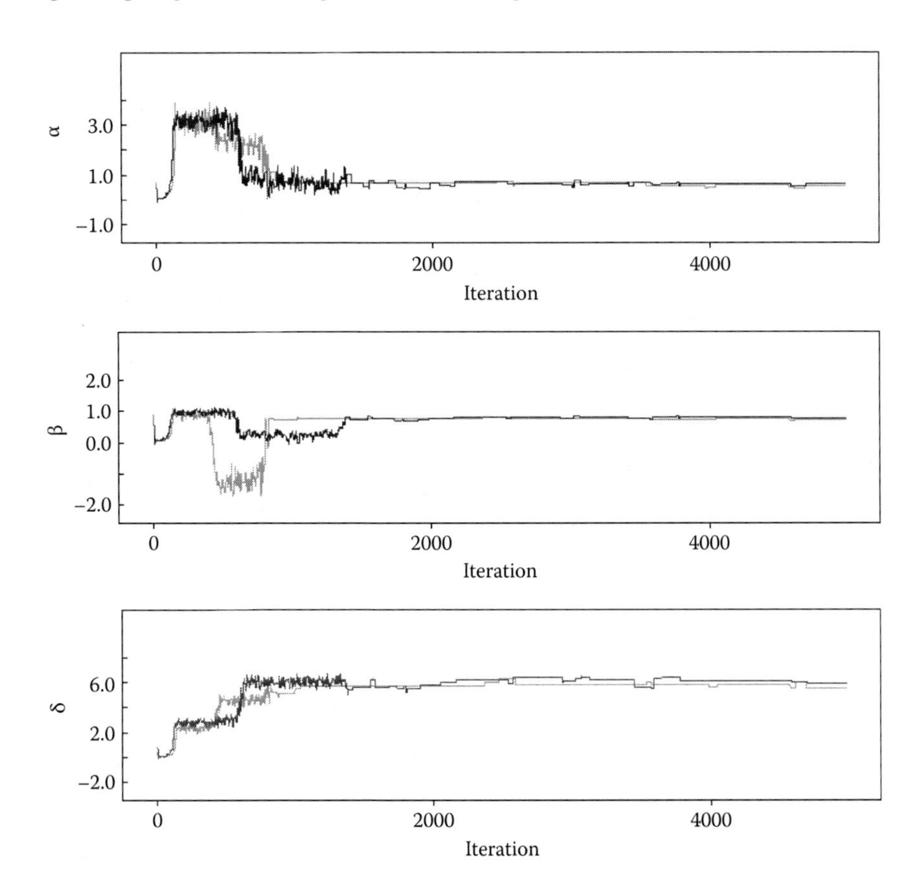

FIGURE 9.8
Final decision tree.

As you can see, the chains converge to a common value after about 1800 iterations, say, 2000 iterations to be safe. This means that the first 2000 iterations are not useful as the system was not yet stable (had not yet converged). So the first 2000 iterations in this case should be dropped off the analysis. This is commonly known as the 'burn in' phase. So from now on, we will keep the *beg* = 2000 and the end to be a very large number beyond the last iteration we have run, say *end* = 10,000,000,000.

We are ready now to obtain a representation for the distribution of the coefficients, alpha (α), beta (β) and delta (δ). First, we run one million iterations (1,000,000). This took about 5 minutes, but depending on the computer capacity you have, it may be faster. Now, we specify again the node we want to visualize at the box *node* on the *sample monitor tool*. Then we hit on the *density* bottom on the *sample monitor tool*. If you forget to change the beginning and ending point, the representation will reflect the issues with the first 2000 iterations. Figure 9.9 illustrates the display of the coefficient before and after removing the non-convergent initial portion of 2000 iterations.

Figure 9.10 shows the distributions of alpha (α), beta (β) and delta (δ). This gives you a good indication of variability and uncertainty. You can also obtain representative statistics for these coefficients; simply select the coefficient you

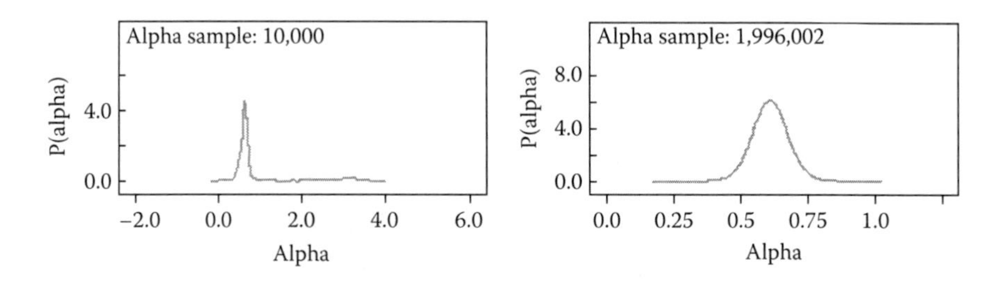

FIGURE 9.9
Before and after removal of convergence.

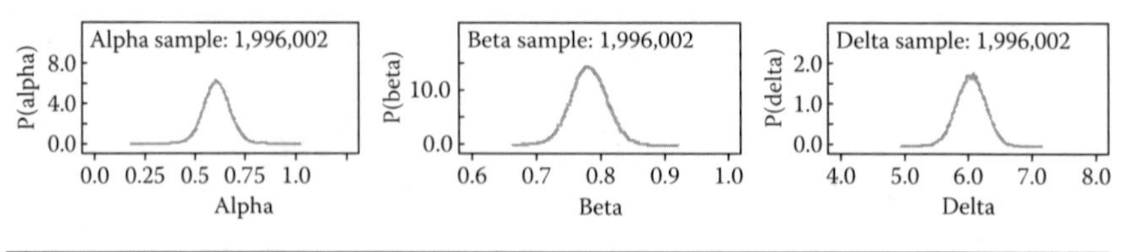

	Mean	Std. dev.	MC_error	val2.5pc	Median	val97.5pc	Start	Sample
Alpha	0.6109	0.06679	4.472E-4	0.4812	0.6104	0.7443	2000	1,996,002
Beta	0.7811	0.02745	2.266E-4	0.7275	0.781	0.8352	2000	1,996,002
Delta	6.061	0.2251	0.001821	5.62	6.062	6.503	2000	1,996,002

FIGURE 9.10
Coefficients.

want to get the statistics for, or, if you prefer use the $*$ on the *node* box, to get them all at once. Hit on *stats* to obtain the statistics.

As a final result, you can see that the mean values match closely those previously obtained in Chapter 5. I can imagine that does not come as a surprise to you because we already specified them before. For this reason, we will change the specification to 'non-informative', meaning that instead of specifying the previously observed values, we will use values of zero as priors and we will use a small value of precision (large variability) and let the system estimate from the data the values for the coefficients. Let's do that: first change on the model the specification of the priors and set the mean to zero and the precision to say -10 to $+10$. That is 10 units away from the mean, so two standard deviations are 10, one standard deviation is 5, and precision is one over the standard deviation squared, that is, $\frac{1}{5^2} = 0.04$ for the value of precision. Second, change the chain values to say -10 and $+10$ so that they depart from very dissimilar numbers; you can choose any other numbers. Figure 9.11 shows the specification of the model once these changes have been made.

We repeat the same steps as before for the specification of the model and loading the data, as well as the specification of the chains. Then we proceed to run iterations. However, soon you will learn that it is more difficult to get smooth convergent distributions for the coefficients.

First, you will see sometimes that running the first iterations may be tricky. So keep hitting on run until the model decides to start running. The problem is that the order of magnitude of the initial values is getting an initial value of the response that is going to negative. Once you escape this trap, run at least

```
# BOOK CESA, chapter 9.

model {
  for (i in 1:20) {
  IRI[i] ~ dnorm(mu[i],tau)  # adapted from Patterson and Attoh-Okine
  mu[i] <- pow(m[i],alpha)*pow(ESALs[i],beta)/pow(SN[i],delta)
  res[i] <- (IRI[i]-mu[i]/sigma)              #standard residual
  p.res[i] <- phi(res[i])                     # accumulated std residual
  Y.pred[i] ~ dnorm(mu[i],tau)
#  P.pred[i]<-step(IRI[i]-Y.pred[i])  # indicator of whether observed > pre
  }
alpha ~ dnorm(0,0.04)                         # Prior expectations informativ
beta ~ dnorm(0,0.04)         Priors           # 2sd = 0.5, i.e. intercept 0.281
delta ~ dnorm(0,0.04)
tau ~ dgamma(0.0001,0.0001)
sigma <- 1/sqrt(tau)
}

list(alpha=-10, beta=-10, delta=-10, tau=0.0001)   Departure
list(alpha=10, beta=10, delta=10, tau=0.0001)      Values
```

FIGURE 9.11
Model specification: Non-informative.

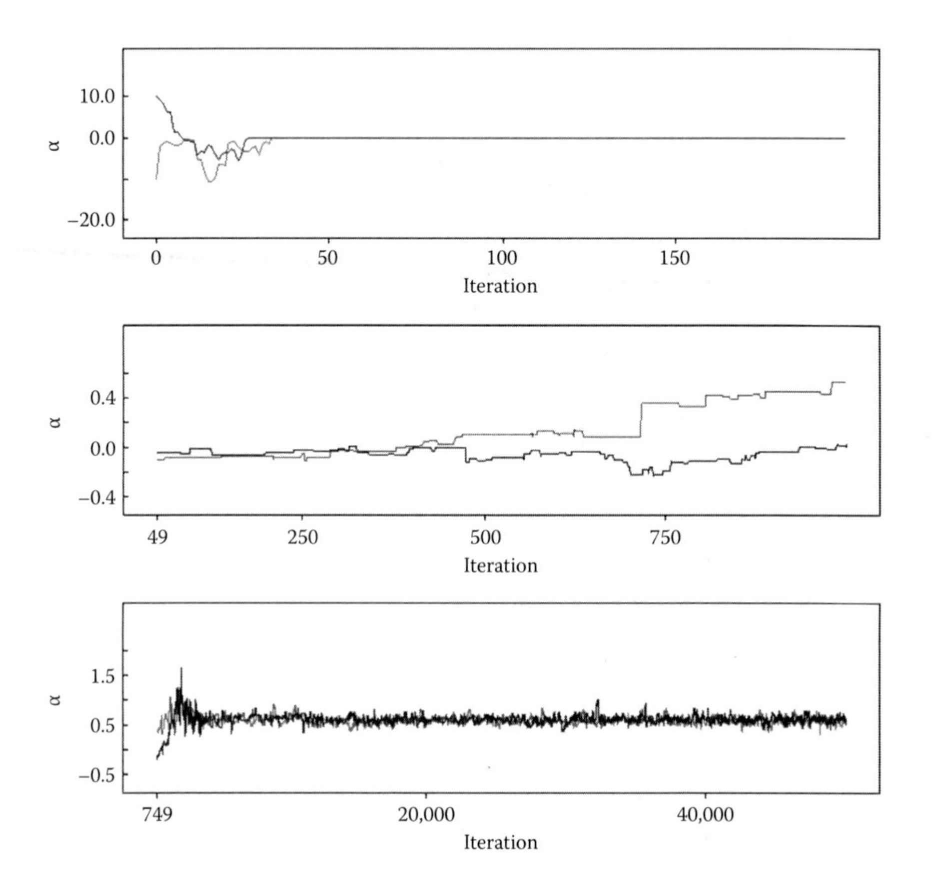

FIGURE 9.12
Model specification: Non-informative.

100,000 iterations and then go and check the history of the chains to check convergence. Let's look at our chains so far. Figure 9.12 shows the chains from three different perspectives: at the top, we can see for the 30 iterations that the chains depart from the initials at -10 and $+10$ and head to a value of 0. After 50 iterations, it appears the chains have converged, however, to zero. When we look at the iterations between 50 and 1000, we can see the values split again after about the 500 iteration. If we look at the third plot, we can see iterations 750–50,000, and we will see that the values have now converged at about 0.6.

So basically we need to drop at least the first 10,000 iterations. We need to change the beginning value to be 10,000, I would recommend. We need to check the convergence of the two other coefficients beta β and delta δ. Run another 900,000 iterations to be on the safe side (you could get coefficients with the leftover 90,000 iterations you currently have).

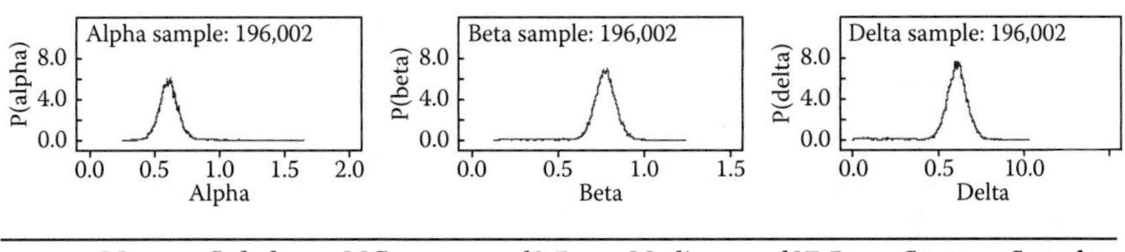

	Mean	Std. dev.	MC_error	val2.5pc	Median	val97.5pc	Start	Sample
Alpha	0.6098	0.07249	4.54E−4	0.4688	0.609	0.7563	10,000	1,980,002
Beta	0.7774	0.06043	3.205E−4	0.6602	0.7768	0.899	10,000	1,980,002
Delta	6.027	0.5629	0.002999	4.937	6.021	7.163	10,000	1,980,002

FIGURE 9.13
Estimation of coefficients: Non-informative.

Previous non-informative estimation confirms the values of the coefficients. I have added the statistics at the bottom of Figure 9.13.

9.5 Uncertainty in Response

Similar to the coefficients, the actual response has been modelled as a probabilistic variable, with a mean value and variability. We will be interested in plotting the 95% confidence interval for the response, that is, we will have two enveloping curves that would give us a graphical representation of how much the response varies. This is an indication of uncertainty.

I will explain the steps to plot the response and obtain the enveloping lines for the 2.5% and 97.5%. The steps are very simple. First, you need to set up a monitor for all the variables you may use. We will do it for *mu* and *Y.pred.* Simply write these elements on the *node* and run your iterations. Second, on the *Inference* tab, select *compare*. The *comparison tool* window will appear (Figure 9.14).

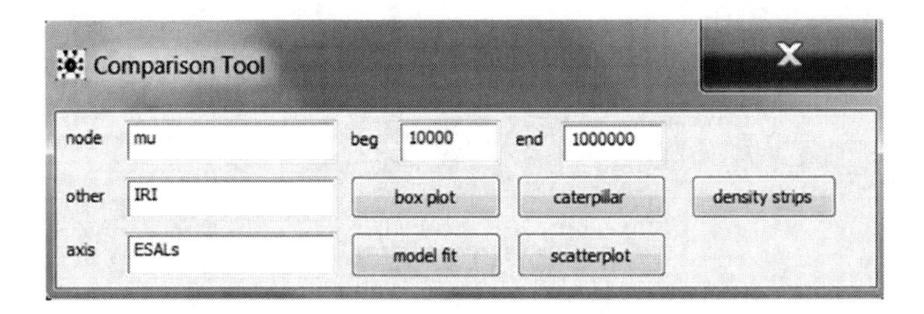

FIGURE 9.14
Comparison tool.

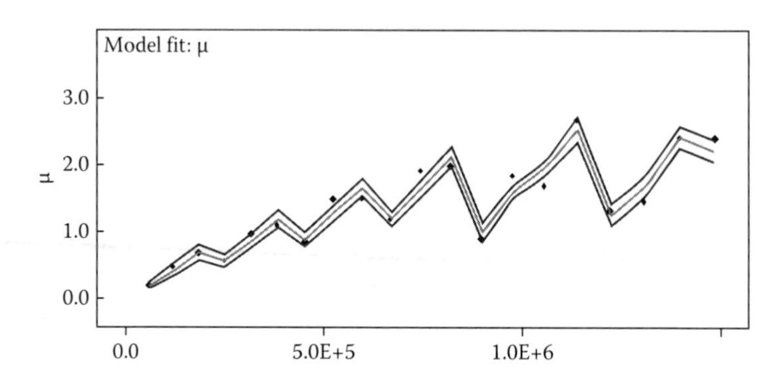

FIGURE 9.15
Model fit: Predicted *IRI* versus *ESALs*.

You will see three empty boxes that need to be defined: node, other and axis. The node reflects the variable y or response (dependent variable). The axis is our dependent variable. Right here you have a problem: there are three variables behind the response (the structural number, the traffic loading in ESALs and the environment in the form of m), and we are only able to plot the effects of one of them (Figure 9.15).

However, as you can expect, the fact that our functional form involves three changing variables will result in a breakdown line because this model explains correlations on one-time year data and not on longitudinal data spanning for several years. This is a very common mistake people commit.

We will make a detour of our attention now to plot a set of data for 20 years of observations. To make it simple, we will have one point per year, but you must bear in mind that on many occasions you may have several points per year. Table 9.8 shows the observations. Notice that all segments were selected from the same environmental region and from the same structure strength. This is a common strategy when working with data; you create a group of segments with very similar characteristics (called 'homogeneous groups').

We repeat here the same process developed before. First, we utilize a functional form that captures the response and the only significant factor (ESALs) that was divided by 100,000. You should notice that the structural number and moisture index were fixed and therefore left out from the model. We chose to use the following exponential form:

$$IRI = \alpha e^{\beta * ESALs}$$

The coefficients to be estimated are alpha (α) and beta (β). We proceed in the same manner as before with the estimation using *OPENBUGS*. The model

TABLE 9.8

Sample Data Deterioration with Fixed m-coefficient

IRI	*ESALs*	*SNC*	*m*
1.61	25,000	3.96	0.04
1.74	50,503	3.96	0.04
1.87	76,260	3.96	0.04
2.00	102,000	3.96	0.04
2.18	128,500	3.96	0.04
2.40	155,000	3.96	0.04
2.30	182,000	3.96	0.04
2.80	209,000	3.96	0.04
2.50	236,000	3.96	0.04
3.20	264,000	3.96	0.04
2.80	292,000	3.96	0.04
3.70	320,000	3.96	0.04
3.30	348,500	3.96	0.04
4.3	377,000	3.96	0.04
3.70	406,000	3.96	0.04
5.10	435,500	3.96	0.04
4.30	465,000	3.96	0.04
6.00	495,000	3.96	0.04
4.70	525,000	3.96	0.04
5.90	555,700	3.96	0.04

set-up is shown in Figure 9.16. You must notice I rounded off the values of ESALs for Table 9.6.

After 200,000 iterations, we obtain a good estimation of both coefficients. The difference now is that they are part of a longitudinal estimation of a functional form, so when we specify a plot of *IRI* versus *ESALs*, the traffic loading reflects the accumulated values through time, and so *IRI* reflects the actual values year by year. Figure 9.17 shows the estimated values of the coefficients.

Alpha takes an average value of 1.216, and beta 0.2953. However, the standard deviation on alpha is much larger than on beta. The 95% confidence interval is shown with blue lines around the mean (red line), and the actual observations are shown as black dots. At about zero *ESALs*, we observe *IRI*'s values of about 1, but as we progress towards 400,000, the value of *IRI* approaches 4, and at 500,000, it approaches 6 (Figure 9.18).

You can alternatively plot the predicted values of the response versus the levels of ESALs. This will result in different lines because it will consider the final estimated value including the variability on the factors.

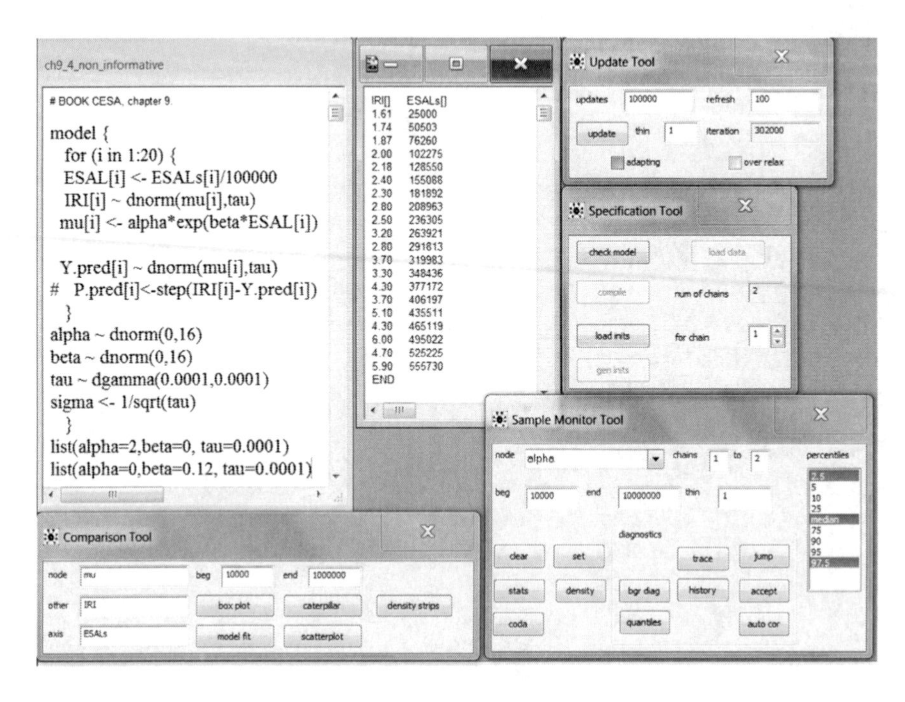

FIGURE 9.16
Model specification.

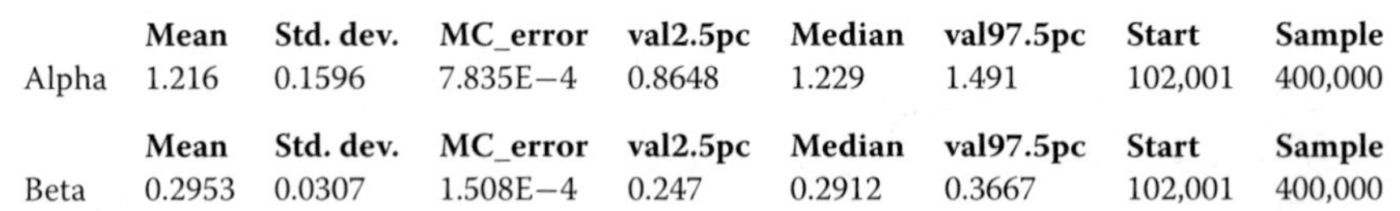

	Mean	Std. dev.	MC_error	val2.5pc	Median	val97.5pc	Start	Sample
Alpha	1.216	0.1596	7.835E−4	0.8648	1.229	1.491	102,001	400,000
	Mean	**Std. dev.**	**MC_error**	**val2.5pc**	**Median**	**val97.5pc**	**Start**	**Sample**
Beta	0.2953	0.0307	1.508E−4	0.247	0.2912	0.3667	102,001	400,000

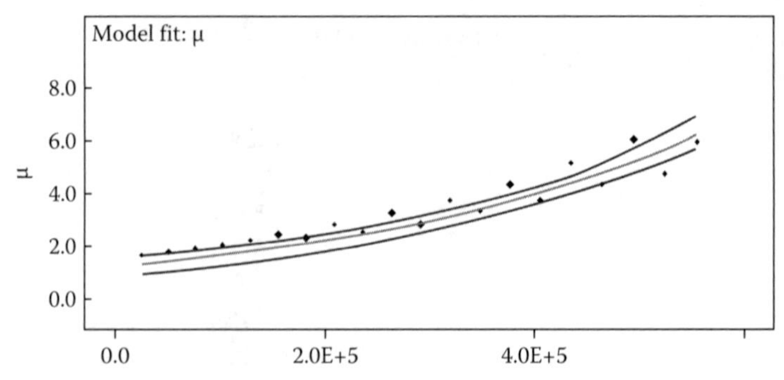

FIGURE 9.17
Model fit: Mean versus *ESALs*.

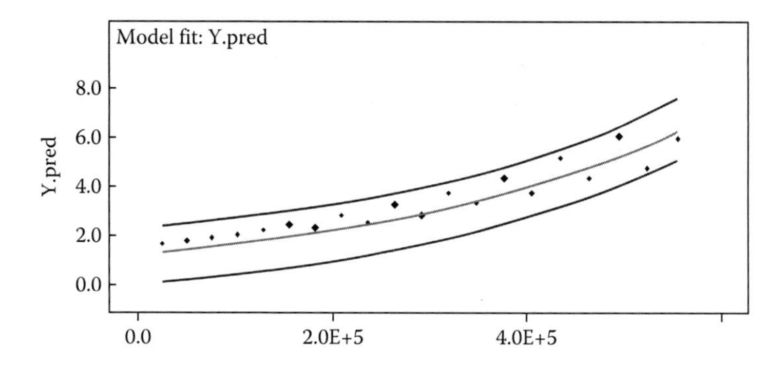

FIGURE 9.18
Model fit: Predicted *IRI* versus *ESALs*.

We can also obtain a box plot of the IRI expression represented by the mean (alternatively you can do it for the predicted response) values along with their variability, as the one we covered in Chapter 4. On the *Comparison Tool*, select *box plot*, and you will obtain a plot as shown in Figure 9.19. Remember that you have prepared thousands of iterations associated with each value of the data set.

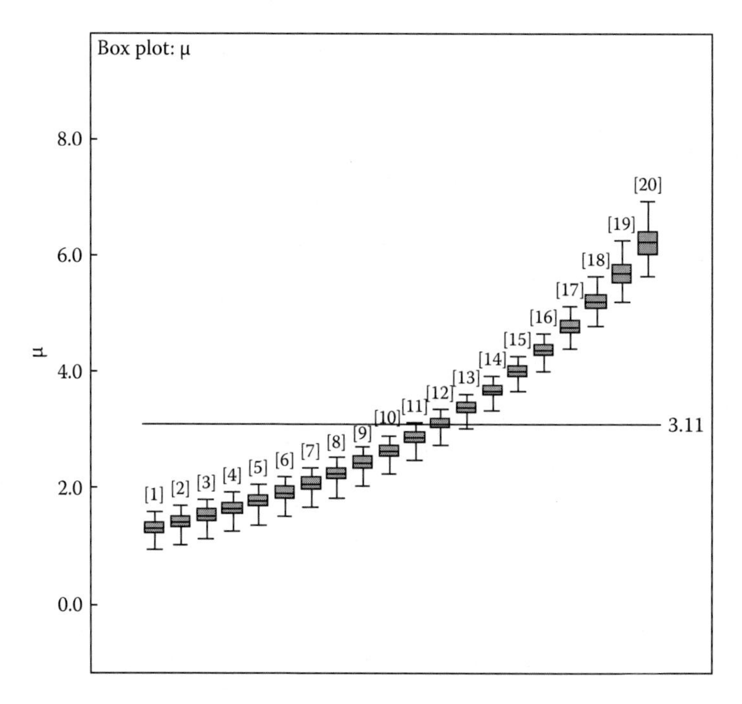

FIGURE 9.19
Box plot for mean (μ) IRI.

TABLE 9.9

Sample Data Road Safety

y	x_0	x_1	x_2	x_3
28	8.78	2	0.5	0
6	8.77	2	0.17	2.67
22	8.78	2	0.5	0
133	9.23	1	0.63	1.13
162	9.13	2	0.8	0.47
2	8.28	2	0	1.80
5	8.28	2	0.17	0.45
23	9.17	2	0.17	2.5
119	9.11	3	0.7	0
162	9.08	3	0.85	0
46	8.27	3	0.43	1.89
3	8.03	2	0	1.5
13	8	2	0	1
8	8.2	2	0	0.45
5	8.2	2	0.33	1
20	8.14	2	0.5	1.4
17	8.3	2	0.5	2.33
2	8.46	2	0.17	2.23
3	8.47	2	0	1.7
3	8	2	0	2
11	8.05	2	0	1.5
3	8.05	2	0	2
12	8.04	2	0.71	0.86
7	6.9	2	0.17	0.65
9	6.9	2	0	0.65
3	7.97	2	0	0
2	7.97	2	0	1.05
2	7.97	2	0	0.53
3	7.78	2	0.33	1.42
4	7.33	2	0	0
2	7.13	2	0.17	1.25
6	7.13	2	0.33	0.93
3	7.14	2	0.33	1.25
2	6.92	2	0	1.2
6	6.23	2	0	0.6
6	6.77	2	0	1.03

Exercises

1. How much uncertainty do you observe? (Table 9.9) The following expression repeats one we covered before in Chapter 5; use it for this exercise. $y = \alpha(x_0) + b_1x_1 + \cdots + b_nx_n$.
2. Using the data presented before, prepare a plot of 95% confidence interval for the response. Also, identify after how many iterations the value of the estimation converges. You can use as initial values those found in Chapter 5. As an additional exercise, you can prepare the same estimation using a non-informative prior by assigning zero mean and large variability to each of the coefficients.

Solutions

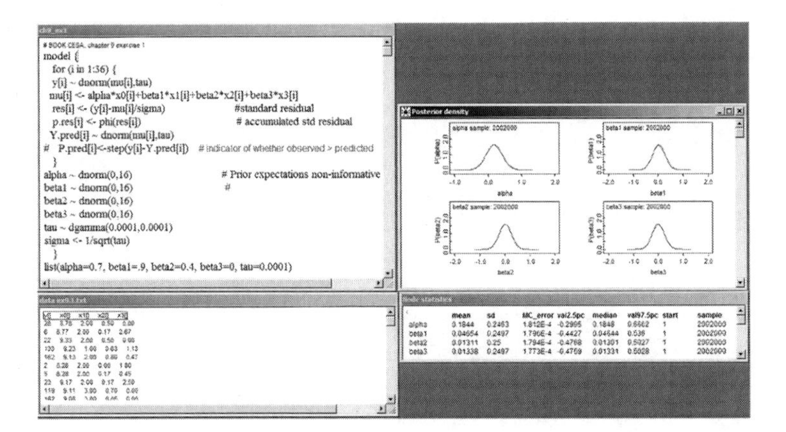

It converges very quickly after 10 or so iterations, the following Figure shows the plots

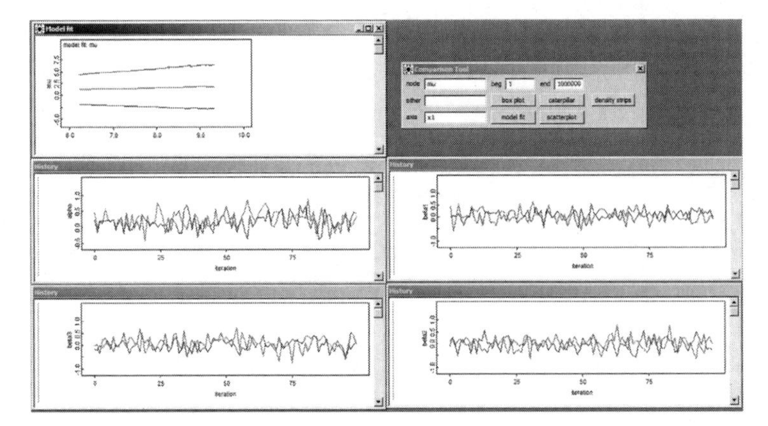

Index